CAUSALITY AND SCIENTIFIC EXPLANATION

VOLUME ONE
Medieval and Early Classical Science

VOLUME TWO
Classical and Contemporary Science

Causality and
Scientific Explanation

William A. Wallace

VOLUME ONE

*Medieval
and Early
Classical Science*

Ann Arbor
The University of Michigan Press

Preface

This work has been long in preparation, too long in fact, and yet even now I am somewhat diffident about presenting it to the scholarly world. For a considerable time I have felt that my work lay outside the main current of that being done in the philosophy of science in the U.S. Recently, however, in light of discussions initiated by Thomas S. Kuhn, Paul K. Feyerabend, and others in what has been interpreted as an attack on the philosophy of science "establishment" in American universities, matters on which I have reflected and lectured for years seem suddenly to have become apposite. To me, for example, it has always been a scandal that the standard expositions of scientific method, usually worked out in a logical positivist framework, have been unable to account for any acquisition of truth in science. According to the strict logic of the hypothetico-deductive method, one could never be sure that the earth is round and spinning on its axis, that the blood circulates, that molecules and atoms exist. And yet most scientists with whom I have worked accept these and like statements as truths; more, they are convinced that the scientific enterprise itself has as its goal the discovery of such truths, and that through their discovery it ultimately contributes to a cumulative growth of knowledge. Apparently many philosophers of science have also shared this belief, even while being unable to justify it through their own logical endeavors. And now Kuhn and others have burst the bubble with what has been characterized as an "irrational" claim that there *is* no growth of knowledge, that science's history may be likened more to a series of Gestalt switches than to any linear progress toward, and ultimate attainment of, truth.

My sympathies, somewhat unexpectedly, are about equally divided between Kuhn and his allies and those who go to make up the philosophy of science "establishment," although at ground I do not agree with either. It has been said of Kuhn that he believes in science but that he does not believe in truth, while Feyerabend, being a realist, believes in truth but he does not believe in science.[1] I too am a realist, and I happen to believe in both. My major concern, like that of all philosophers, is how to substantiate such a twofold belief. In advancing this goal I have sought to enrich the philosophy of science literature with a more searching examination of scientific explanation than had hitherto been available, particularly of the paradigm of scientific explanation that has perdured through so many centuries of science's history, namely, causal explanation. Such a search has led me to studies in the history of science, and so, like Kuhn, I see history as making an essential contribution to the philosophy of science, although our works issue in markedly different results.

My hesitancy in publishing my findings even at this stage stems precisely from my knowledge of historical studies. Perforce I have gone back to Aristotle for my beginnings, and I have had to work out in some detail the progress of science over the eight centuries that stretch from 1200 to the present. No one man can claim expertise over such a long period of history, nor can one even control the major primary and secondary sources. I make no pretense on this account to bibliographical completeness, let alone exhaustiveness, although I have striven to provide enough documentation to enable the reader to retrace my steps. Because my thesis is unusual, if not controversial, I have attempted also to supply as many citations of primary sources as possible, thereby allowing the reader to make his own interpretation of them rather than forcing on him my own. For this reason the work may be use-

---

[1] H. R. Post, "Correspondence, Invariance, and Heuristics: In Praise of Conservative Induction," *Studies in History and Philosophy of Science*, 2 (1971), p. 253.

ful as a source book even for those who disagree with its general argument. As to the argument itself, I feel that the time is ripe for formulating it and inviting the criticism of my peers. A work of synthesis cannot wait forever; there will always be more facts to uncover, but at some point the scholar must have the courage to attempt to pull together what he knows, even at the risk of making mistakes of omission and interpretation. I feel that I have reached such a point, and in doing so that I may have acquired enough humility to accept both the possibility of error and the help that others may give to enable me to reach the truth.

The work was originally planned as a single volume treating of causality and scientific explanation in three, successive, chronological periods, those namely of medieval science, classical science, and contemporary science. As writing progressed, however, it became apparent that the final version would have to be divided into two parts, and that the dividing point would fall in the classical period. Thus the work now appears in two volumes, the first dealing with medieval and early classical science, the second with classical and contemporary science. This division offers the advantage that the reader can study, in one volume, the elements of methodological continuity between medieval and early classical science, and, in the other, the factors that caused contemporary science to emerge from the downfall of classical science while still searching for a methodology that would justify its results. The thesis advanced throughout both volumes, however, is that the reconstruction of contemporary philosophy of science necessitates a return to some of the values of medieval and early classical science, and thus for philosophical integrity the volumes should be considered as a unit. From the viewpoint of history of science, on the other hand, they are relatively independent and should be able to stand on their own merits.

All citations from primary sources I have given in English, from standard translations when available, otherwise in my own translation. Generally in citing early British authors I have modernized English spelling and punctuation, and not infrequently I have modified or adapted existing translations on the basis of my own reading of the text; in matters of any substance I have called attention to the latter in a note.

I have debts too numerous to acknowledge in detail, particularly to colleagues, students, friends, and the librarians who have assisted me; all of these I ask to regard the work itself as my best expression of thanks. I wish explicitly to acknowledge, however, a research grant from the National Science Foundation that enabled me to read extensively in American and European libraries during the academic years 1965–66 and 1966–67. It was this opportunity that put me in contact with the many original sources on which my studies are based, and without which they could never have been undertaken.

William A. Wallace

January 28, 1972
Washington, D.C.

Contents

VOLUME ONE

CHAPTER 1. Prolegomena: Explanations and Causes   1
  1. The Contemporary Problem   1
  2. Method of Approach   7
  3. The Legacy of Antiquity   10
    a. Aristotle   11
    b. Pythagoras and Plato   18
  4. Later Development   21

Part I. Medieval Science

CHAPTER 2. Medieval Science at Oxford   27
  1. Grosseteste and His Influence   28
    a. Roger Bacon   47
    b. John Peckham   51
  2. The Mertonians   53
    a. William of Ockham   53
    b. Walter Burley   55
    c. Thomas Bradwardine   58
    d. Dumbleton and Heytesbury   59
    e. Richard Swineshead   60
  3. The Oxford Contribution   62

CHAPTER 3. Medieval Science at Paris   65
    1. Albert the Great   66
    2. Thomas Aquinas   71
        a. Mathematics and Proof   80
        b. Mathematics and Hypothesis   86
    3. Experimental Method   88
        a. Peter of Maricourt   88
        b. Theodoric of Freiberg   94
    4. The Paris Terminists   103
        a. Jean Buridan   104
        b. Albert of Saxony   109
        c. Nicole Oresme   111
    5. The Paris Contribution   114

CHAPTER 4. Padua and the Renaissance   117
    1. Early Paduans   118
        a. Paul of Venice   121
        b. Gaetano da Thiene   127
    2. Revival at Paris   130
        a. Jean Mair   131
        b. Dullaert and Coronel   132
        c. Celaya and Soto   135
    3. Sixteenth-Century Padua   139
        a. Agostino Nifo   139
        b. Jacopo Zabarella   144
        c. Experimental Method   149
        d. Copernicus and Hypotheses   151

Part II. Early Classical Science

CHAPTER 5. The Founders of Classical Science   159
    1. William Gilbert   162
    2. Johannes Kepler   168
    3. Galileo Galilei   176
        a. The Early De motu   177
        b. Galileo's Methodology   180
    4. William Harvey   184
        a. Aristotelian Demonstration   185
        b. Causal Explanations   190

5. Isaac Newton 194
    a. The Experimentum crucis 196
    b. The Cause of Gravity 205
Notes 211
Bibliography 251
Index 269

CHAPTER ONE

Prolegomena:
Explanations and Causes

1. The Contemporary Problem

WHEN in 1961 Ernest Nagel came to publish his definitive treatment of scientific methodology under the title, *The Structure of Science*, he added a significant subtitle, *Problems in the Logic of Scientific Explanation*.[1] Of particular interest in the latter is his use of the word "explanation," for, contrary to a view that has a long history in studies of scientific methodology, Nagel insists that the aim of science is not merely to describe, to give answers to the question "How?," but ultimately to explain, to give answers to the question "Why?" Thus he openly defends the thesis that "the distinctive aim of the scientific enterprise is to provide systematic and responsibly supported explanations," with explanations being understood as answers to the question "Why?"[2] In taking this stance, Nagel has opted for a view of science that has been gaining currency among philosophers of science, particularly those who have grown dissatisfied with positivism and are inclining now toward one or another type of realism.

The fluid situation in contemporary philosophy of science that Nagel's position reflects is one reason for interest in the problem of scientific explanation. Another reason, which reflects a more critical attitude toward the role of explanation in science, may be found in the state of science itself in the present day, which is undergoing such rapid change that one may question whether any of its explanations can be enduringly valid. Yet another is the esoteric character of many of the theoretical explanations proposed by scientists, which makes them practically unintelligible to the non-specialist, so much so that one may wonder precisely as to their explanatory value.

With regard to the changing state of science, it is undeniable that science, as a major part of the information explosion, is in a period of rapid growth and has had to assimilate changes at an unprecedented rate. There is always uncertainty during periods of rapid change, and one of the concerns of scientists, and of teachers of science, is whether explanations that are being given today will be regarded as valid tomorrow. The question may be asked whether modern science is appreciably different in this respect from medieval alchemy or astrology. During the Middle Ages it was quite common to explain chemical changes in terms of only four elements. Is there any assurance that the elements of modern chemistry will not face obsolescence in much the same way as did these four? Similarly, in astronomy, are the insights provided by the general theory of relativity basically any different from those of the Ptolemaic system with its elegant equants, eccentrics, deferents, and epicycles? This is a legitimate concern of theorists who would address themselves to the problems of modern science. Science is in a period of flux; principles and facts that at one time were regarded as fundamental are being called into question. Do the explanatory systems that link them belong ultimately in the same category as the speculations of the Middle Ages or of antiquity? If so, can a scientific explanation be enduringly valid? [3]

The esoteric character of the explanations currently accepted among scientists may be seen in two theories that have changed twentieth-century science in a fundamental way, quantum theory and the theory of relativity. The first, as is well known, finds its principal application in the region of the very small, the microcosm, while the second relates mainly to the domain of the very large, the megalocosm. In both areas man is quite remote from sense experience and from the ordinary ways of verifying his knowledge. The entities discussed in quantum theory are too small to be observed with even ideal optical instruments; they are said to be unobservable in principle. Likewise, many of the phenomena discussed in relativistic cosmologies are too remote to observe or even to identify without considerable theoretical extrapolation. In the region of the very small a variety of constructs have been found necessary to explain the experimental data of atomic and nuclear physics. An example might be the wave-particle construct, according to which matter may be said to have both a wave structure and a particle structure. This manner of speaking has resulted in concern with problems of consistency, not to say of overt contradiction. In maintaining that the same reality is both a wave and a particle, is not one endowing it with what — on the basis of ordinary language and common experience — can only be regarded as contradictory characteristics? In the region of the very large similar questions have been posed about space and time and the relationships between them. At an earlier period in science's history the concept of simultaneity was regarded as completely meaningful and acceptable; it was implicitly agreed that one could know when two events were simultaneous, even when they occurred in different places. With Einstein's formulation of the relativity principle, and its accent on procedures of measurement through the transmission of light signals, so-called paradoxes have suddenly appeared.[4] There seems to be no way in which absolute simultaneity can be defined operationally and in unambiguous terms. Heisenberg's statement of the uncertainty relation in quantum mechanics has led to similar difficulties there.[5] For centuries bodies were thought to be specifically determined and exactly located in space and time. In nineteenth-century thought, science itself was regarded as providing the most determinate knowledge one could hope for. The universe was regarded as a machine, completely determined in structure and operation; it seemed to be run by causal laws, to operate with clockwork precision. Now, however, there is a suspicion that perhaps matter itself is ultimately undetermined, that a basic indeterminacy attends all the operations of nature. In the face of fundamental revisions in conceptions such as these, is it any longer possible to give explanations for the events of common experience in terms that make sense to an intelligent but non-specialized observer?

Closely allied to the foregoing, and yet another reason for the growth of interest in scientific explanation, is a concern with the work that has been done since the turn of the twentieth century in the philosophy of logic and of mathematics. Much of this was prompted by the discoveries of non-Euclidean geometries and by new theories of number, without whose help the relativity and quantum theories could never have been elaborated. Attempts to formalize these discoveries led to serious study of the foundations of geometry and of arithmetic. One result is that the notion of axiom, long accepted as self-evident truth, has had to yield to the weaker notions of postulate, rule, and definition. The concern with truth and certainty has given way to a search for consistency and the elimination of contradiction from formal systems. In the consequent burgeoning of mathematical logic, categorical argumentation has gradually been replaced by conditional or postulational reasoning, mainly on the basis of material implication. The resulting orientation in logic has done much to solve problems in the foundations of mathematics, but it also prompts questions of a methodological nature relating to the natural sciences. If formal logic can clear up fundamental problems relating to explanation, if truth tables and matrices are adequate to the task of analyzing scientific methodology, one is tempted to ask whether scientific explanation is itself nothing more than logical explanation. When analyzing the findings of modern science, can one hope solely for the discovery of logical antecedents from which the data of science can be derived? Or, to put the question somewhat differently, given an *explanandum* or *explicandum*, can one hope to find an *explanans* or *explicans* that goes beyond the domain of logic?

The position that all explanations in science are nothing more than logical explanations has been advanced by an influential school in the philosophy of science, that of logical positivism. Its proponents have held that science is limited to collecting positive data and then linking these data together in formal systems, wherein logic provides the connections and enables one to make predictions of new data. The ultimate antecedents in such systems are theoretical entities that are postulated by the scientist; they exist in his mind as logical entities, and need not be thought of as having extra-mental existence.[6]

Now, if all scientific explanation is merely logical explanation, the claims of realism become much attenuated in the scientific enterprise. In particular, one wonders if there can ever be explanation in terms of real entities. If the antecedent in terms of which one explains the data of modern science is merely a conjectured or fictive entity, the fabrication of the mind alone, what relation does this have to objective reality? Should the explanations of modern science lead only to fictional entities, to *entia rationis*, their value for understanding the world of ordinary experience is minimal. The scientist may appear to have explanations for many phenomena of nature, but when asked directly, "What is this thing in terms of which you purportedly explain?" he seems able to reply only that it is something he made up, something that exists in his mind and possibly there alone.

This general situation is one about which philosophers and scientists have been much concerned, and rightly so. The practicing scientist, in particular, is committed to the real world; he believes that he is attaining to knowledge of reality, whether he states so explicitly or not. Thus the question of the value of explanation in modern science, and particularly its real or ontological import, is for him of considerable interest.

So much for explanation. Coupled with this concern over explanation is the recurrence, not so frequent but nonetheless noticeable, of questions concerning causality.[7] One can talk to contemporary scientists or read their books and find little reference to causes or to causality; yet, those who are concerned with problems of explanation, and particularly with the question, "What is a causal explanation?" have found it necessary to discuss this vexatious topic. The topic is vexatious because it has a history, and a history that has witnessed serious oppositions on the basis of philosophical commitment. The names of Hume and Kant among others bring to mind differences that have polarized discussions in a fundamental way.

Going back to the medieval precursors of modern science, one finds that they consistently associated the Latin term *scientia* with causal knowledge. *Scientia est cognitio per causas.* It was precisely on this basis that they differentiated *scientia* from other forms of knowledge: science was concerned with causes, whereas other types of knowing were not. To have scientific knowledge was to know perfectly, *scire simpliciter*, to know that something is so because it could not be otherwise and this, in turn, because of the causes that make it be as it is. Thus, the early understanding of science — from which the modern notion grew — was that it must be concerned with a search for causes. And the explanations for which science was ultimately searching, from its beginnings among the Greeks to the seventeenth century, were causal explanations.

The characterization of science as causal knowledge thus has a long tradition. It perdured through much of the eighteenth and nineteenth centuries, with various alterations dictated by changes in philosophical viewpoint. Among the latter, those promoted by empiricism and positivism gradually came to have the decisive influence. One of the major tenets of empiricism was that man cannot come to a knowledge of causes; positivism, in its turn, insisted that all that can be known are positive data and correlations among them. As these philosophies gained in influence, science came to be disassociated from the traditional search for causes. In the view of scientific activity that came to prevail in the early twentieth century, there is no way in which the scientist can penetrate beneath appearances to detect any ontological influx or causal influence. In place of causal explanation, therefore, he must seek logical connections or functional correlations as the proper goal of his science.

This positivist view of science has been dominant in the Western world throughout the twentieth century. Recently, however, under the impetus of studies in the history of science that reveal the continuing interest of prominent scientists in problems of epistemology and even of metaphysics,

there has been a renewal of interest in causality as it relates
to scientific explanation. Some of the reasons for this renewal
are to be found in dissatisfactions with the Humean analysis
of causality and with the excessively formalistic analyses that
have dominated the literature in the philosophy of science. A
more basic concern is the issue of realism in modern science,
for this reopens the question whether science can attain to explanations that provide not only the logical antecedents of
phenomena but also their ontological antecedents. An ontological, or extra-mental, antecedent of phenomena would actually be equivalent to a cause. This, at any rate, is a traditional meaning of the term "cause": it is something that exists
outside the mind and serves to explain, not merely in a logical but in an ontological way, why the thing is as it is.[8]

2. *Method of Approach*

Hence the association of causality and scientific explanation.
The problem raised by this association is of contemporary interest in the philosophy of science, and relates to science not
only as it has been practiced in the past but as it is being
practiced in the 1970's and will be practiced in the future.
Despite the importance of the problem, however, its solution
is by no means easy, nor is agreement to be expected even on
methods of approach. Here there are two obvious possibilities, corresponding to the two directions being taken by the
philosophy of science in the present day, namely, the analytical and the historical. The analytical approach would be
problem-oriented, based essentially on problems recognized
as such in twentieth-century science, and consisting in detailed and formal analysis of the causal concept as related to
explanation that is commonly regarded as scientific. This
mode of analysis could be extended to include the various
meanings of the term "cause" in ordinary language, and the
relevance of this usage to scientific explanation. The historical
approach, as opposed to this, would proceed mainly by a
study of the history of the usage of the term, illustrating how
it has functioned in various periods of the evolution of science. It in turn could be extended to include case studies of

classical contributions in different areas of investigation, with comparative analyses to show the recurrence of various patterns of explanation throughout the history of science.

Most of the recent writing on scientific explanation has been in the analytical context, and this by way of offering more or less definitive solutions to the problem that would eliminate or severely curtail the role of causes or causal explanations in modern science.[9] Little if any attention has been given to the historical study, in depth, of the relationships between causes and explanations, particularly in such a way as to leave latitude for a more benign appreciation of the values of causal explanation for solving problems in contemporary philosophy of science. To fill the lacuna that thus exists, the second approach will be favored in this study, although not so exclusively as to ignore or fail to take into account recent writings on the subject of scientific explanation in the analytical mode.

Several arguments may be adduced in support of this procedure. The first is that causality and explanation were already linked in Greek and medieval science, whence modern science took its origins. In early science, as already noted, to give an explanation was to give a causal explanation. As science matured, however, the connection was gradually abandoned, and presumably for serious reasons. An historical study can perhaps uncover these reasons and open them to philosophical analysis. Certainly anyone who would wish to restore the association between causality and explanation would be well advised to know why it had fallen into dissuetude, what difficulties had been encountered in the past, and what insights might encourage future attempts at its restoration.

A second reason is that the technical terminology of contemporary science does not lend itself readily to analysis in realist terms. Recent explanations of gravitation, for example, employ complex mathematical functions developed within postulational or operationalist frameworks, whose ontological import on this account is extremely difficult to disengage. It is here that the vantage point of history can provide a perspective and a point of reference that is more open to realistic analysis and interpretation. Science has proceeded generally along the same path as all human knowledge, which goes from ordinary things, from macroscopic objects near at hand, to those that are more remote from experience such as the very small and the very large. Man proceeds from the knowledge of what he sees to the knowledge of what he cannot see or of what may be perceptible to him only through the use of instruments. Generally, too, he uses ordinary language to describe things that are near at hand and employs a minimum of technical terminology to account for them. There is no problem, for example, in understanding the realism involved in the statement, "A body falls," particularly when this is located in the context of Galileo's studies of falling bodies at Pisa and Padua. It is not nearly so easy to see how theoretical constructs, such as space-time geodesics, provide a realistic explanation for that phenomenon.

Yet another reason for emphasizing the historical over the analytical approach is the value of historical studies for freeing the mind from unconscious commitments and presuppositions. History is a great equalizer, no less in the world of ideas than in the affairs of men. Sooner or later it reduces all schools and movements to their proper size, not infrequently by showing how quickly the popular can become the ephemeral. This liberating influence of history should be especially welcome in areas where one approach has tended to dominate discussion of a problem, such as the point of focus in the present study.[10]

The emphasis on an historical approach, however, should not obscure the fact that the problem being addressed pertains to the philosophy of science. On substantive issues historical analysis can lay the groundwork for answers, but the answers themselves must still be given in the context of philosophy. This point being made, it would obviously be presumptuous to attempt to resolve all questions relating to causality that are being discussed in contemporary philosophical literature on the basis of historical analysis alone. Such is not the intent of the present study. It aims rather at providing a new setting in terms of which current philosophical problems can be discussed, with the expectation that an understanding of how causality and explanation were linked in men's minds for over two millennia can only be helpful for advancing the discussion.

3. The Legacy of Antiquity

These considerations naturally lead to an exposition, in the light of history, of the causal terminology to be employed. In recent years several serious studies have been made of the origins of both mathematical and experimental methodologies that have come to characterize modern science. Working back through the "precursors of Galileo" in a variety of ways, these studies converge on work done in the early thirteenth century at Oxford University and at the University of Paris.[11] This work, in turn, had its origins when the *Posterior Analytics* of Aristotle, which had been lost for centuries or had existed only in fragmentary or unintelligible form, was finally rediscovered by the Latin West. The Aristotle who thus came to be known was not a pure Aristotle, for the text had passed through one or more translations. When first read, moreover, it was interpreted in a thought context that included Platonic elements, among others, and was appraised critically mainly from the viewpoint of its conformity to, and usefulness in, current systematizations of Christian theology.

Such extraneous influences aside, the *Posterior Analytics* seems to have provided the inspiration from which the initial medieval contributions that later centuries would identify as scientific actually derived. Robert Grosseteste, Albert the Great, and Roger Bacon, to mention only the principals, were among the first to comment on the newly available text of the *Posterior Analytics*, apparently with the intention of putting its canons of scientific method into practice.[12] After them, a consistent methodological tradition developed that continued all the way to Galileo, Harvey, and others who are rightly regarded as the founders of modern science. Thus it appears that this treatise and its commentaries were central factors in formulating scientific practice. Other Aristotelian treatises, such as the *Physics* and its commentaries, were subsidiary factors in that they served to illustrate how the procedures outlined in the *Posterior Analytics* could be put to work in the physical sciences themselves. For this reason, in what follows a brief summary will be given of Aristotle's doctrine on causality and explanation as contained mainly in the *Analytics* and the *Physics*. Then, in view of the Platonic currents that merged with Aristotle's thought as this was transmitted to the Latin West, the elements of Platonic and related doctrines on causes and explanation will be treated. This should suffice for documenting the legacy of Greek science that bears on the terminology employed in later scientific methodology. Where the Latin Aristotle diverges appreciably from the Greek Aristotle, and also for purposes of subsequent reference, notes will be appended to the exposition to illustrate the influence of translators on the understanding of Aristotle's thought to the end of the twelfth century. The distinctive interpretations put on these teachings by thirteenth-century and later thinkers will be left for fuller treatment in subsequent chapters.

a. Aristotle

The *Analytics* of Aristotle are divided into two parts, the *Prior Analytics*, corresponding to what is generally called formal logic, and the *Posterior Analytics*, corresponding to what is less frequently called material (in contradistinction to formal) logic. The latter is concerned with the theory of demonstration or proof, probably the most important element in Aristotle's scientific methodology. The superior kind of knowing that he calls science (*epistēmē*) can only be achieved by one who can successfully identify the cause (*aitia*) that makes a fact (*pragma*) be what it is: he must know the cause from which the fact results, as the cause of that fact and no other, and accordingly that the fact cannot be otherwise.[13] Causes thus function for Aristotle as reasons or explanatory factors that make scientific knowing possible. The whole of the second book of the *Posterior Analytics* is concerned with explicating how it is possible to give such explanations, and the precise way in which causes will enter into their formulation. To proceed scientifically, for Aristotle, is essentially a matter of putting questions to nature, and his detailed analysis centers around the way in which such questions may be asked. Only four basic possibilities are recognized by him, and these are listed rather cryptically as questions of fact (*to hoti*), questions of reason (*to dioti*), questions of existence (*ei esti*), and questions of nature or essence (*ti estin*).[14] Using a teaching he has already developed in the *Prior Analytics*, according to which the middle term of the categorical syllogism gives the reason or explanation why a predicate may be applied to the subject under investigation, Aristotle reasons that all four of these questions are related to this middle term. The first and third questions, in his analysis, really inquire whether there is an explanation or a "middle" (or, more radically, whether *any* explanation is possible), whereas the second and fourth questions, presupposing that an explanation or "middle" can be sought, inquire precisely what this is. Moreover, since the explanation will show cause why the predicate can be attributed to the subject, the middle term will have to reveal the cause, and thus all four scientific questions are intimately connected with the concept of causality.[15] It is this line of reasoning that leads Aristotle to the conclusion that causality is inseparably linked with explanation, understanding the latter in the sense of a middle term.[16]

Within the field of scientific knowing Aristotle makes a distinction between "knowledge of the fact" (*to hoti, quia*) and "knowledge of the reasoned fact" (*to dioti, propter quid*).[17] Both are answers to scientific questions, but the second yields the more scientific knowing since it provides not only knowledge of the fact but an explanation as to why the fact is as it is. Yet it sometimes happens, Aristotle notes, that a reasoning process leading only to knowledge of a fact indirectly supplies an explanation from which knowledge of a reasoned fact may be obtained. When this is the case, a logical rearrangement of the explanatory factors and predicates involved can convert *quia* knowledge to that which is *propter quid*. One of Aristotle's examples, of interest for its possible applications in astronomy, is the proof that the planets are near because they do not twinkle. So stated, the explanation results in knowledge of a fact, provided one has already established, through induction (*di' epagōgēs*) or through sense perception (*di' aisthēseōs*), that whatever does not twinkle is near. This explanation, however, does not provide knowledge of a reasoned fact, since "not twinkling" is not the cause of, or the reason for, the planets' "being near," even though it can generate an awareness of their proximity. Should one see, on the other hand, that the nearness of the planets is the explanation or proximate cause of their not twinkling, he can rearrange the terms of the argument so as to obtain knowledge of a reasoned fact, namely, that planets do not twinkle because they are near. Another example that illustrates the same possibility is the proof that the moon is spherical from its phases, where one's awareness that the moon exhibits phases leads to the knowledge of the fact that it is spherical, whereas the sphericity of the moon, once known, provides the proximate causal explanation of its exhibiting phases.[18]

Yet more interesting examples of the differences between knowledge of the fact and knowledge of the reasoned fact are given by Aristotle when he discusses cases where mathematical reasoning is applied to physical problems. In such cases, Aristotle notes that it is for the collector of data, or empirical observer (*aisthētikos*), to know the fact and for the mathematician (*mathēmatikos*) to know the reasoned fact, since the latter is in possession of demonstrations giving causes.[19] Aristotle here provides two examples, the first of which relates to scientific knowledge of the rainbow, where he maintains that it is for the natural scientist (*phusikos*) to know the fact of the rainbow's existence, whereas it is for the optician (*optikos*), whether simply such or as a mathematical optician, to know the reasoned fact. The other example concerns the relationship between medicine and geometry with respect to the fact that circular wounds heal more slowly than others. Here it is for the doctor to know the fact, whereas it is for the geometer to know the reason for the fact, although Aristotle does not state what that reason is.[20]

Knowledge of the reasoned fact, for Aristotle, is evidently the goal or summit of scientific explanation, and as such is related in one way or another to causal explanation. He recognizes four different types of causes, which were labeled formal, material, efficient, and final causes by the medievals, though this terminology is not so explicit in Aristotle. In the *Posterior Analytics* he enumerates the four causes as follows: [1] the definable form (*to ti ēn einai*), [2] an antecedent which necessitates a consequent (*to tinōn ontōn anagkē tout' einai*), which may also be translated as the necessitating conditions, or ground, in the sense of a matter or substrate, [3] the efficient cause which started the process (*hē ti prōton ekinēse*), and [4] the final cause (*to tinos heneka*).²¹ This enumeration is repeated in the *Physics*, together with series of examples that serve to illustrate the ways in which the different types of causes may be understood.²² [1] As examples of the formal *aitia*, or definable form, Aristotle gives the form (*eidos*) or pattern (*paradeigma*), the reason or definition (*logos*), either in a specific or a generic sense, and the whole (*holon*) or synthesis (*sunthesis*) that results from the union of various parts. Yet other examples are the idea (*idea*), the form (*morphē*), and the shape (*schēma*) that is characteristic of the thing. [2] For the material *aitia*, or necessary substrate, he mentions that out of which (*ex hou*), as a constituent, something is generated, and instances bronze with reference to a statue and silver to a cup. In other contexts he refers to this as the matter (*hulē*) and the subject (*hupokeimenon*), and derivatively as an element (*stoicheion*) or a part (*meros*) as it goes to make up an integral whole. [3] The efficient *aitia* is the primary source (*archē*) of change or of coming to rest, and is exemplified by an adviser, a father, any maker or agent (*hothen*), and again by a seed and a physician, all of which initiate the process of a change or its cessation. [4] The final *aitia*, lastly, is the purpose or end (*telos*) for which a thing is done, as walking (or reducing, or purging, or taking medicines) for the sake of one's health. Other usages are that for the sake of which (*charin*), as inquiry for the sake of knowledge, and the good (*agathon*) as the final and best possible achievement.²³

Apart from enumerating the four types of causes, Aristotle is concerned with showing how they may be related to each other precisely as explanatory factors. He is insistent that the physicist or natural philosopher (*phusikos*) must seek a comprehensive answer to the question "why?" and that in so doing he will uncover explanations in terms of all four factors. Sometimes the same reality will count as an explanatory factor under different headings, depending on the point of view being analyzed. Again, in certain kinds of investigation, explanations are more readily given in terms of one type than another. Yet the most comprehensive answer to the "why" question will be in terms of all four types of explanatory factors, if these can be attained.[24]

This summary presentation of Aristotle's treatment of causes as explanatory factors in science suggests several observations relating to current misconceptions of his teaching. For one, modern authors, looking for a simple illustration of what is actually a complex doctrine, frequently hit upon what may be referred to as the statue analysis. According to this, the material cause is the marble out of which a statue is made, the formal cause the figure or shape of the statue, the efficient cause the chiseling or other action of the sculptor, and the final cause the purpose for which the statue is intended by the sculptor. Such an example, while making good sense of the terminology, leads to a superficial understanding of what Aristotle intended by causal analysis. In the first place, artifacts are such obvious and trivial cases that their explanatory factors need hardly be inquired into, whereas natural entities are frequently refractory to understanding, and require analysis in terms of various explanatory factors precisely to be understood. Again, the completeness of the statue analysis creates the impression that it is possible, or even necessary, to identify all four causes in any given analysis of nature. It is Aristotle's practice, on the contrary, to employ only as many types of explanation as are appropriate to the matter in hand. It is possibly for this reason that he rarely gives the same example to illustrate a number of causes, but rather changes his examples each time when describing the types of explanatory factors. The statue analysis, again, confers a spurious equality on all four types of cause, and makes it difficult to understand how, in opposition to the material cause, the formal, efficient, and final causes frequently coincide, and how there are very special relationships that connect pairs of causes, such as material-formal and efficient-final. Finally, the over-simple identification of formal cause with figure or shape does nothing to assist the understanding as to how the formal cause, for Aristotle, is the essence or quiddity of the thing as expressible in its definition.[25]

The statue analysis, however, does serve to illustrate a device that is characteristic of Aristotle, namely, that of extrapolating from cases that are understandable on the basis of ordinary experience and developing therefrom a technical terminology that is homogeneous with, and intelligible in terms of, the vocabulary used to deal with familiar cases. His explanatory paradigms differ on this account from many employed in modern science, which are based on idealized or limiting cases that are not encountered in simple observation or in ordinary discourse. Aristotle's doctrine of causes illustrates rather well this concern with linguistic and conceptual foundations. His treatment of matter, or material causality, starts with cases where the underlying substrate is easily seen, but then is extended to cases that are less easily understood in terms of simple observation, until finally it arrives at a protomatter, spoken of by the Latins as *materia prima*, that in some respects is as sophisticated as the concepts of modern field theory. Much the same can be said for his analysis of final causality as teleological explanation, for he starts with examples of purposive activity that are readily seen in human affairs, and then moves by successive stages, usually discarding conditions inherent in the situations from which he starts, to discern comparable purposiveness in the behavior of plants and animals and in the functioning of their various parts.[26]

Aristotle's success in extrapolating from the familiar case is attested by the fact that much of his terminology is still incorporated in modern scientific discourse. The concept of mass, for example, is not Aristotelian, and yet its definition as "quantity of matter" utilizes two concepts that received their first formulation from Aristotle. Similarly the use of such terms as energy (*energeia*) and potential (*dunamis*, whence the term "dynamic"), with their wide applications in all branches of modern science, is traceable to his careful handling of these terms so as to describe great numbers of situations with a common terminology. Other instances are the discrimination of the chance (*tuchē*) and random event (*automaton*) from the more obvious cases of causal determination, the technical formulation of problems relating to the continuum, and precision in the use of many terms that have become part and parcel of discussions on scientific methodology, such as induction, deduction, definition, and demonstration. In the more logical treatises such as the *Categories*, Aristotle has also bequeathed to the sciences a host of terms that are indispensible for classification, coordination, and subordination. The term "category" itself, as well as genus, species, difference, property, and accident (in the sense of an accidental feature, as opposed to something essential) are all part of Aristotle's heritage.[27]

Also noteworthy is the systematic attempt by Aristotle to apply his scientific methodology to the whole range of beings found in the material universe. Contrary to a frequently expressed opinion, he was not intent on biological explanation alone, but worked out general principles that would be applicable to nature's entire domain, even allowing for the use of mathematics — and this particularly in the treatment of the non-living — but all the while insisting that the search for proper principles would lead to distinctive explanations at all levels, including the sciences of man. Throughout this development, it is true that the methodology of the *Posterior Analytics* is rarely referred to, nor is it conspicuously in evidence. This work, however, was never proposed by Aristotle as a model for inquiry, but rather as a development of canons that would be necessary for rigorous exposition once such inquiry had been completed.[28]

Against those who would maintain that Aristotle wished to eliminate mathematics from physics, it is necessary to insist that mathematical proof constituted Aristotle's first paradigm for scientific explanation, as opposed to Plato's theory with its stress on dialectics. Again, for Aristotle mathematics was a science concerned with the physical world, although at a higher level of abstraction than the physical sciences, and moreover it was not essentially a science of transcendent objects, as it had become in Plato's view. This explains Aristotle's continued use of mathematical argumentation throughout the *Physics, De Caelo,* and the *Meteorology.*[29] The latter work is particularly interesting for its detailed treatment of the rainbow and other radiant phenomena, which were among the first physical phenomena to be subjected to mathematical investigation. It is in this work too that Aristotle shows himself aware of the role of falsification in the growth of scientific knowledge, a role only recently accorded primacy in methodology by Karl Popper.[30] Discussing explanations proposed by his predecessors for such phenomena as comets and the Milky Way, Aristotle shows how they must be rejected or revised because their consequences are disproved by observational evidence. "Though more could be said," he concludes one exposition, "this is enough to demonstrate the falsity of current theories of the causes of comets." [31] The *Meteorology* is the locus also for Aristotle's pointed remark that "the final cause is least obvious where matter predominates," and at the conclusion of the entire work he states simply that "we know the cause and nature of the thing when we understand either the material or formal factor in its generation and destruction, or best of all if we know both, and also its efficient cause." [32] Significantly he makes no mention here of final causality as a source of explanation in meteorology — a science in his system of classification that probably has more in common with modern physical science than any other discipline.

b. Pythagoras and Plato

Since mathematical explanation is differently treated by Aristotle and by his predecessors, notably Pythagoras and Plato, it will not be amiss here to give a summary exposition of the Pythagorean and Platonic accounts. [33] Pythagoras and his school were among the first to attempt to understand the cosmos and its phenomena in terms of number, although they built on foundations that had already been laid by the Egyptians in geometry and by the Babylonians in arithmetic. Pythagoras's investigation of the properties of integral numbers and the possibilities of arranging numerical sets (such as triangular numbers) in geometrical form suggested a parallelism between the idealizations of geometry and the physical patterns evident in the universe. Stimulated by this insight he was led to maintain that number underlies all physical objects and that its study reveals a deeper level of reality than is apparent on the surface. Aristotle reports that the Pythagoreans considered number to be the principle not only of the matter of things but also of their form, with the elements of number being the even or the "unlimited" which functions as the matter, and the odd or the "limited" which serves as the form. [34] In their view, whatever the nature of primordial mat-

ter it must always take on the form of number and consequently of geometrical shape, with the result that every physical entity is ultimately explainable in terms of mathematics.[35]

From this basic insight the Pythagoreans were led to a
detailed study of mathematical proportions and harmonic relationships and ultimately to the discovery of irrational numbers. The latter was of special importance because of the
stimulus it gave to studies of the continuum and of the paradoxes relating to infinite divisibility. Thus Pythagoras and his
school fostered a deep appreciation of the role of mathematics in the study of the physical universe, and exerted a considerable influence on Plato and the mathematicians of the
Hellenistic period, Archimedes included; this was to perdure
through the centuries and would ultimately be revived in the
seventeenth century in the thought of Galileo and Kepler.

Plato made use of Pythagorean ideas, combining these
with Democritus's atomism to offer a geometric theory of
matter. He regarded matter as a stable and eternal receptacle
for Ideas or Forms, and spoke of it as "the mother of all
becoming." [36] Material entities, the objects that fall under the
observation of the senses, are not the most real beings nor are
they even subsistent in themselves. Since the physical world
is a changing, phenomenal world, in order to come to know
reality one must slough off the appearances, the shadowy
forms of material things, to apprehend the Forms that lie behind appearances. Ultimate reality, as Plato saw it, is to be
found only in the world of autonomous, immaterial Forms;
the formless matter out of which the shadowy appearances
were made may have been, for him, an ultimate factor, but it
is not reality itself. Moreover, if matter is the mother, the
eternal Form is the father, and the transitory phenomena
known by the senses are their offspring. The eternal Ideas
cast images on a formless matter that serves as a receptacle
and as a dimensional space (*chōra*) in which the images can
be generated, so that these same images can be said to be
made up of a form (*morphē*), which is an individual copy of
the subsistent Form, and a matter (*hulē*), whatever stuff it is
that goes to make them up.[37]

Coupled with this teaching on matter and form, Plato
saw each eternal Idea as embodying a perfect organization

that could be expressed in logical definition, which itself was
ultimately based on a geometrical and a numerical structure.
He thus conceived the elements of the physical world as constructed on the model of geometrical patterns, with fire taking the shape of a tetrahedron, air an octahedron, and water
an icosahedron. All of these shapes, moreover, are multiples of
isosceles triangles, so that the triangle functions as an
"atomic" element somewhat in the Democritean sense. In the
Timaeus Plato sought to explain how the elements could be
generated from triangles and from each other, obviously on
the supposition that a basic mathematical structure underlay
the physical world.[38] Perhaps for this reason in his Academy
the study of mathematics was regarded as the key that would
open the door to nature and to wisdom.[39]

Yet for Plato true wisdom consisted in the contemplation
of eternal Forms and not in the study of the transitory shadows that make up the physical world. The study of numbers
as pure Forms yields truth, in his view, but this is not the
same as studying the Pythagorean numbers that inhere in
concrete objects and are impure on this account. Even astronomy, which studies the most perfect of material things, has
for its object the visible world and thus falls far short of attaining to true realities.[40] Plato is thus critical of the attempts
of the Pythagoreans to discern the underlying mathematical
structure of reality from experiments and from the study of the
metrical properties of objects. Physics, in his view, could be
only a "likely story," and thus he did not share the epistemic
ideal of a science of nature that was to be formulated by his
celebrated student, Aristotle.[41]

Although differing in their manners of conceiving the
role of numbers as causes and the use of mathematics in scientific explanation, neither Pythagoras nor Plato thought of
mathematics, as did Aristotle, as a formal and somewhat abstract discipline that could be applied to physical reality and
thus generate the subalternated type of scientific knowing described in the *Posterior Analytics*. Rather both opted for a
process whereby one would go through the appearances of
reality to a knowledge of mathematical form, as the ultimate
and basic reality itself. For both Pythagoreans and Platonists,
therefore, mathematics would not be applied to nature so

much as discovered beneath nature's appearances, and from this underlying strategic base provide a more or less reliable pattern for scientific explanation.[42]

Closely associated with the less realistic Platonic evaluation of mathematical explanation is the methodological device of using mathematical theories "to save the appearances" (*sōzein ta phainomena*), particularly the appearances of the heavens associated with the movements of the heavenly bodies. The expression itself seems to have been first used by Simplicius, a sixth-century commentator on Aristotle, who mentions in his commentary on the *De Caelo* that Plato posed the following problem for the mathematicians: "What circular motions, uniform and perfectly regular, are to be admitted as hypotheses so that it might be possible to save the appearances presented by the planets?"[43] The device of using hypothetical motions and other constructs to "save the appearances" has a long history following its seeming introduction in this text of Simplicius. Some interpreted it in a strictly Platonic sense, to mean that a physical or real explanation of the heavenly bodies could never be arrived at, and that one must settle as a consequence for a mathematical or formalistic accounting as the ultimate attainment of which science is capable. Others understood the expression to mean that provisional mathematical theories should be sought as interim devices for prediction and calculation, but that they need not replace physical explanations along more realist lines, once sufficient data became available for their consistent formulation. Obviously the history of these alternate conceptions of scientific method bears closely on the point of this study, and will have to be taken into account in its subsequent development.

4. Later Development

These considerations lead naturally to a sketch of the further historical development and thus to the outline of the present work. Since the subject is being approached historically, it may be conveniently divided into three parts, the first concerned mainly with medieval science, the second with classical science, and the third with contemporary science. The treatment of medieval science will have to build on the concepts of Greek science, some of which have already been sketched, but which were subjected to searching examination and revision throughout the high medieval and late medieval periods, extending all the way to the Renaissance and the sixteenth century. The main concern throughout will be to examine critically what scientific contributions were made by the medievals, and how causal analysis and other forms of scientific explanation functioned in their discovery and subsequent justification. As will be seen, no single or monolithic method was employed throughout this period, although different lines of investigation came to characterize the more important university centers. Oxford University was the scene of the pioneering efforts, and those working at Oxford generally placed great emphasis on mathematical explanations, at least partially because of the Platonist-Aristotelian atmosphere then present at the university. The University of Paris, on the other hand, started with a more empirical orientation because of the purer form of Aristotelianism that evolved there, but it gradually assimilated currents from Oxford, particularly in the fourteenth century, and gave rise to a view of science that had considerable affinity with that accepted in the early modern period. At the University of Padua, finally, the methodologies developed at Oxford and Paris were discussed in settings that were more open to experimentation in the modern sense, and thus provided the proximate setting for the emergence of the "new science" with Galileo as its principal proponent. The treatment of medieval and Renaissance science will therefore be centered around the three universities of Oxford, Paris, and Padua, as symbolizing various components that entered into the "new science," while recognizing that there was considerable interplay between these three centers of learning during the entire period, and that the history of scientific ideas cannot be geographically localized in a way that the sequence of treatment may seem to suggest.

In examining the second period, which extends roughly from the beginning of the seventeenth to late into the nineteenth century, two considerations will have to be borne in mind. First, the early modern period, for all the protestations of reaction against late scholastic and Renaissance methods, was dominated by the same search for causes and thus was in recognizable methodological continuity with the medieval period. This is apparent not only in the work of the founders of modern science but also in their frequently expressed reflections on logical procedures. Secondly, early modern science triggered in large measure the development of classical philosophy, and this in turn had considerable feedback on the science that influenced its own growth, particularly in matters methodological, reaching all the way to the present day. These two factors suggest a threefold division of the classical period, the first focusing on the founders of modern science, the second on the philosophers who dealt with the new science they produced, and the third with the methodologists who were influenced by both scientists and philosophers and attempted to formulate explicitly the canons of investigation that characterized science in its classical understanding. Since the first of these treatments may be more easily assimilated to Greek and medieval thought, it will serve to bring to a close the matter discussed in the first volume. The latter two treatments, on the other hand, serve well to introduce the problematic for contemporary philosophy of science, and thus pertain more to the matter of the second volume. All three, however, are still treatments of classical science, and a common concern dominates their discussion. Throughout, attention will be centered on the role of causality in scientific explanation, and the various meanings that were attributed to causes and to causal explanations. It will be seen that causality was never completely relinquished as a source of scientific explanation from the seventeenth to the mid-nineteenth century, although many were dissatisfied with the causes that had been proposed by their predecessors and in their place sought to expound the "true causes" of physical phenomena. Each successive formulation seems to have introduced a slight change in the meaning of causality, however, until finally the notion of causal explanation was rather completely linked with determinism and predictability along quite mechanistic lines.

The third period comprises developments that took place in the latter part of the nineteenth and throughout the whole of the twentieth century. It is this period that saw the downfall of classical science and the origins of modern physics with the relativity and quantum theories. The deterministic concept of causality was, of course, rejected along with the science that embodied it, and a continuing concern with the problem of what to substitute in its place has given rise to what is now recognized as the philosophy of science movement. An attempt will be made to locate discussions on causality and scientific explanations with respect to the various schools that characterize this movement, and from this to adjudicate some of the claims in favor of causality that have emerged within the last two decades. What will result from this is a reassessment of the role of causality in scientific explanations, with a proposal for the reinstatement of causal explanation along fuller lines than characterized its classical usage. This may prove helpful for serious reconstruction within the philosophy of science, particularly with regard to the ontological value of explanations in general, and thus for a renewed interest in realism within the scientific enterprise.

Part One
Medieval Science

CHAPTER TWO

Medieval Science at Oxford

WHETHER one begins an account of medieval science with Oxford or with Paris, to say nothing of the Italian universities, the choice will be to some extent arbitrary. If attention is focused on the more mathematical of the physical sciences, however, and if moreover the chronology of a particular school is traced, there are convincing arguments for placing the beginning at Oxford University. It was there that Robert Grosseteste first suggested a strong relationship between mathematical and physical explanation, and it was there, in comparative isolation from Rome and from a variety of pressures on the Continent, that he was able to form a cohesive school that worked on common scientific problems.[1] Grosseteste's influence, in fact, perdured at Oxford throughout the thirteenth century and reached well into the fourteenth, where it was still to be found in the refined mathematical physics developed at Merton College. Accordingly, our point of departure for discussing causality and scientific explanation in the Middle Ages will be Oxford University, and the analysis will center on two phases, the first that of Grosseteste and those he influenced and the second that of the group known as the Mertonians.

1. *Grosseteste and His Influence*

"The esteemed Robert, onetime Bishop of Lincoln, of holy memory, took no notice at all of Aristotle's books and their methods," wrote Roger Bacon, "but from his own experience and other writers and from other sciences he got involved in Aristotle's inquiries, and he found out and wrote more than a hundred thousand times as many things as are discussed in Aristotle's books or can be gotten out of the perverse translations of the same." [2] An interesting statement, if true, but like many that come from Roger's pen, in need of substantial qualification. For Grosseteste did read other authors and he did make original investigations, but he also read Aristotle seriously and pondered over his text, studying not only its findings but also its methodology. The "perverse translations" did not always help matters, but they nonetheless stimulated this first master of the Oxford Franciscans, and from the marriage of corrupted text with creative mind came the birth of mathematical physics in the Latin West.

Thus, at least, goes a commonly accepted account, following Crombie and others.[3] The account has much to commend it, for Grosseteste was not only an assiduous translator of Aristotle and other Greek authors but he also commented carefully on the *Physics* and the *Posterior Analytics*.[4] Anyone who reads these pioneering if sometimes frustrating efforts at exegesis cannot help being amused at Roger Bacon's comment, nor can he be other than impressed with the extent of Grosseteste's own contribution to the founding of science as we now know it.

In bare outline, the essentials of that contribution have been detailed as follows. Reared in an atmosphere of Augustinian theology and with a strong personal bias for Neoplatonism, Grosseteste subscribed to the then-current doctrines at Oxford of exemplarism and emanationism, of divine illumination and light metaphysics. The latter, in particular, inclined him to believe that the universe was created and structured by the auto-diffusion of light in geometrical patterns, and his attitude toward science was therefore based on this ontology. When he read Aristotle's account of science in the

Posterior Analytics, and particularly when he saw there the distinction between science *quia* and *propter quid,* he found the key he had been looking for. Physics by itself could be science *quia,* providing knowledge "of the fact," but aided by mathematics it could also become science *propter quid,* yielding knowledge "of the reasoned fact." In either event, of course, science must consist in a search for causes through effects, and generally this could be effected by resolving (i.e., by analysis, hence the need of the *Analytics*) complex phenomena to simple principles, by either a process of definition or a process of demonstration. When principles and causes had been attained, they could then be composed in the order of their production to yield an explanation for the desired effect. Hence the basic method of science was that of resolution and composition. As employed in mathematics and metaphysics, the method yielded perfect accuracy and certitude, but as employed in physics, even with the aid of mathematics — and now a Platonic conviction asserted itself — it could yield only probability. Here a further step became necessary: explanations in physical science had to be checked, they must be subjected to a process of verification and falsification. From this last step was thus born the concepts of experimental science, for experiments, in thought if not in deed, are required to verify and falsify alternative physical explanations. And so Grosseteste and his followers joined experiment (in principle, at least) to mathematics, and in so doing made a firm beginning that was to culminate, four centuries later, in the "new science" of Galileo.[5]

To the bones of this skeleton, of course, should be added the actual contributions of Grosseteste and Bacon and Peckham to all of medieval science, but particularly to optics and meteorology, where rudimentary mathematics could be used in discovering nature's secrets. Such documentation may be postponed for the moment; it will show quite clearly that the Oxford ideal of science in the thirteenth century was explanation through causes, preferably through causes that are somehow quantifiable or amenable to mathematical treatment.

This general account has an air of plausibility about it, and yet, to anyone who peruses with care Grosseteste's commentaries on Aristotle's *Physics* and *Posterior Analytics,* it does not ring completely true. The account does accord well, no doubt, with Grosseteste's opening statements in *De lineis*, of which the following are representative:

> There is the greatest utility in considering lines, angles, and figures, for without them it is impossible to know natural philosophy; they have absolute value throughout the entire universe and its parts. . . . For all causes of natural effects are brought about through lines, angles, and figures. Otherwise it would be quite impossible to have a *propter quid* science of them.[6]

These sentences express a very clear and forthright position, quite consistent with a Neoplatonic view of the universe. And yet there are difficulties latent in the account as a whole. For if one can make physics or natural philosophy into a *propter quid* science through the use of mathematics, why maintain that its explanations never yield certitude? Why make the further connection with verifiability and falsifiability, and urge this as a somewhat anachronistic motivation for the use of experimental method in the Middle Ages? Perhaps Grosseteste's concept of physical science is not as consistent as it might be; perhaps, on the other hand, the Oxford master is actually more nuanced than his interpreters. Not an easy matter to decide, but let us turn for help to the commentary on the *Posterior Analytics* and study the physical examples Grosseteste uses there to illustrate the Aristotelian ideal of science, and particularly the relationships that hold between sciences *quia* and *propter quid*.[7]

The classical *locus* for discussion of the differences between demonstrations *quia* and *propter quid*, as has been noted, is chapter 13 of the first book, and at this point in his commentary Grosseteste makes an interesting observation. He states that up to this chapter Aristotle has been discussing science and demonstration in a perfect and restrictive sense, but that now he will discourse generically on these topics to include the way in which they are applicable to physics or natural philosophy.[8] This observation leads him into a classification of the two types of demonstration, namely, through a proximate cause (*propter quid*) and not through a proximate cause (*quia*), the latter being also referred to as demonstration *per posterius.* He observes too that there can be *propter quid* and *quia* demonstrations of the same conclusions, and that sometimes these demonstrations will pertain to different sciences, called respectively the subalternating science and the subalternated science, while at other times they will pertain to the same science.[9] Thereupon he gives examples drawn mainly from astronomy that will be treated shortly.

From the way in which Grosseteste starts out this section, one might interpret him to mean that Aristotle has discussed perfect or *propter quid* demonstrations in the first twelve chapters, but that now he has finished with this subject and is turning to *quia* or *per posterius* demonstration as generic and applicable even to physics. Another interpretation, however, and one more consistent with the total context is to understand Grosseteste to mean that Aristotle has expanded the discussion of demonstration to now include the generic types, while continuing to discuss perfect demonstration and how it might be related to the generic. The first interpretation would exclude the possibility of *propter quid* demonstration in physics, whereas the second would allow such a possibility, while admitting also the use in physics of the *quia* types.

Under either interpretation, moreover, it would seem that Grosseteste is maintaining that true demonstration can be attained in physics, and thus that its conclusions are certain and not merely probable, as they would be if *dialectica* rather than *scientia* were there involved. This effectively would rule out the need for verification or falsification, unless this be regarded as itself part of the demonstrative process.

That the second interpretation is the more likely may be reinforced by an earlier discussion in Grosseteste's commentary on chapter 8, where Aristotle makes the statement that the conclusion of a perfect demonstration "must be eternal" and that "therefore no attribute can be demonstrated nor known by strictly scientific knowledge to inhere in perishable things."[10] An obvious example is the eclipse, a transitory phenomenon, and Aristotle questions whether this can be the subject of strict, *propter quid,* demonstration. Grosseteste apparently believes that eclipses, although "perishable," can be such a subject, but he likewise feels that eclipses must somehow be universals and incorruptibles, despite their being of intermittent occurrence. So his problem reduces to this: How can universals be incorruptible when their singulars are corruptible, since universals obviously depend on singulars for their existence whereas singulars depend on universals for their demonstration? This strange problem of universals he attempts to solve in typical Neoplatonic fashion by admitting four classes of incorruptible universals: (1) those contemplated in the First Cause, the uncreated exemplars of all that is to be caused; (2) those seen in the created light of intelligence; (3) causal reasons of terrestrial species whose individuals are corruptible, seen in the powers or virtues of the heavenly bodies; and (4) the formal causes of such species in Aristotle's understanding of these as true quiddities from which demonstrations can be made, although a weak intellect can know such quiddities only through their accidents, which thus function for it as principles of knowing and not of being. The last two classes of universals are troublesome for Grosseteste, since their individuals are obviously corruptible; the only further solutions he can offer for saving their universals are (a) that the reasons or forms are not corruptible of themselves, though their deferents are corruptible, or (b) that such reasons or forms, though susceptible to accidental corruption, are *de facto* preserved through a continuous succession of individuals.

Applying this complex list of possibilities to the case of the lunar eclipse,[11] Grosseteste finds difficulty with the very last named, viz. (b), since lunar eclipses are obviously not going on "every hour," and thus a universal eclipse based on such individuals is not really incorruptible. The only other possibility he can see is to retreat to the third type of incorruptible universal sketched above, viz. (3), and maintain that the universal eclipse always exists in its causes or causal reasons, whether or not any individual eclipse actually exists at any moment. But Grosseteste does not give up even at this; he puzzles again over why Aristotle is so sure that eclipses can be the subject of demonstration, and he discovers one final alternative. What Aristotle could have meant, he says, is that the eclipse demonstration is just one step away from being verified at any time, *est proximo habens veritatem in omni hora,* in this sense: it specifies the exact conditions under which an eclipse will occur at any time, and the moment such conditions are verified, the demonstration itself is true.[12] Thus Aristotle's meaning is that events of frequent occurrence are demonstrable, so long as they are specified under the very conditions that will make them true any time the conditions are fulfilled.

This, after all, is a surprising statement for a Neoplatonist who would be expected to live in a world of unchanging essences. The astronomical knowledge behind it is not exceptional; it is all contained in Grosseteste's *De sphaera,* similar in many respects to the work of Sacrobosco with the same title, and based on the eternal, regular, and seemingly predictable rotations of the heavenly spheres.[13] In such an astronomy the causal reasons of eclipse are incorruptible enough to yield true universals, and therefore a science of eclipse. Yet Grosseteste is willing to admit here, with Aristotle, the additional possibility of *conditioned* universals generating a science of frequent occurrences, and this would seem to open the way to a science of corruptible things even in the sublunary sphere.[14]

Recall now that what is being discussed is a *propter quid* demonstration of a corruptible thing, an eclipse. Is such *propter quid* demonstration, in Grosseteste's view, ever attainable in science understood in this generic sense, e.g., in physics or natural philosophy? To answer this, let us return now to the development of the commentary at chapter 13 of Aristotle's text, where Grosseteste explains first the relationship between demonstration *quia* and *propter quid* in the same science, and here he gives as examples the locations of planets and the shape of the moon, and then the same relationship in different sciences, where the examples he analyzes are the rainbow and the healing of a wound.

Limiting discussion first to the subject matter of one science, Grosseteste states that if cause and effect happen to be convertible and if the cause should be more known to a particular observer, this observer can have a *propter quid* demonstration of the effect; if, on the other hand, he knows the effect alone from sense observation, the best he can attain is *quia* demonstration of the cause.[15] For example, anyone who is sure "by astronomical demonstration" that the planets are near, knows *propter quid* that they do not twinkle, just as anyone who is certain "by natural demonstration" that the moon is spherical, knows *propter quid* why it has phases. On the other hand, the person who lacks astronomical science can still observe that planets do not twinkle and from this he can demonstrate that they are near, but this knowledge is "through an effect" and therefore attained by *quia* demonstration. The same applies to the man who does not know why the moon is a sphere; he can reason from its phases, an effect of its sphericity, to their proper cause, and he will then know that the moon is a sphere, but again his demonstration is only *quia*.[16]

At this point Grosseteste raises the interesting possibility of their being, in any one science, both *propter quid* and *quia* demonstration "of the same conclusion." [17] Notice above that in each case the two types of demonstration are of different conclusions, relating to a phenomenon's cause and effect respectively. Can there be both a *propter quid* and a *quia* demonstration of a cause (or of an effect) in the same science? Grosseteste's answer seems to be perhaps, "although Aristotle does not give an example of this." [18] He thereupon enters into a lengthy excursus apparently aimed at showing how the two cases just discussed, viz., the moon's sphericity and the planets' nearness, are susceptible to demonstration both *quia* and *propter quid*. The only difficulty with his ensuing exposition is that it is hard to see how, in so doing, he manages to stay in each case within the domain of a single science. Let us consider now the two examples in some detail, beginning with that of the planets' non-twinkling as a sign of their nearness.

Following Euclid, Ptolemy, and Alkindi, Grosseteste explained human vision in terms of a qualified extramission theory, wherein visual rays go forth from the eye to the object seen, although the object also emits radiation and has some effect on the eye.[19] This theory committed him to analyzing optical phenomena in terms of a visual cone or pyramid, whose apex is at the eye and whose base is the object seen; when an object is far away, the angle at the apex is small, whereas when near, the angle is large. The eye, on this theory, tries to discern the fine structure of the objects it sees, and uses its "visual power" to do so. When the angle of vision is large, it has no difficulty generating sufficient power and as a result it sees a stable object; when the angle is small, however, it experiences difficulty and "generates a tremor in the visual spirits." As a result, the object it sees in the latter case undergoes "minute and rapid tremors," and this apparent variation in a luminous object is called twinkling. Since planets do not twinkle, they subtend a large angle at the eye (compared to stars), and thus they are near. So we have a *quia* demonstration of their nearness.[20] Regarding the *propter quid* demonstration, Grosseteste gives no details in this section of his commentary, stating merely that the planets' nearness is known *per demonstrationem astronomicam*.[21] If we regard astronomy in the peripatetic tradition as one of the sciences of the *quadrivium,* that, namely, determining the positions of the heavenly bodies from geometrical figures generated by the passage of light rays, we can say that the two demonstrations pertain to a single science in the sense of a geometrical optics.[22] This is not the only interpretation possible, however, particularly if we concentrate on Grosseteste's Neoplatonism and take literally some of his statements in his treatise *De luce*.[23] Grosseteste did hold, as already noted, that the entire universe was structured by the self-diffusion of light, and he may have been convinced that he could prove that the planets took up their positions at creation in a determinate mathematical way, through a more fundamental physical optics consonant with his "metaphysics of light." [24] If so, the brief phrase *per demonstrationem astronomicam,* in the broader context of his entire philosophy, could mean that optics, considered alternately in its more geometrical and physical aspects, can supply two demonstrations, *quia* and *propter quid* respectively, of the nearness of the planets. This is not an Aristotelian teaching, of course, but Grosseteste has already admitted this and is attempting to go beyond the text on which he is commenting.

The case of the moon's sphericity is treated by Grosseteste in similar fashion.[25] His *quia* demonstration, based on the generally accepted geometry of the *De sphaera,* again involves the visual pyramid theory of sight, and the analysis is carried out in terms of the pyramid subtended at the eye by the moon's surface and another, larger pyramid whose base is the curve of the illumined portion of the moon, which curve he calls a circle "*causa brevitatis.*"[26] His proof consists in showing that the moon's phases can effectively be accounted for by the rotation of the "circle" which is the base of this larger pyramid around the moon's disc (itself the base of the smaller pyramid), and that granted the conditions of the moon's illumination by the sun in conjunction, quadrature, and opposition, the moon must have a spherical shape to produce the observed appearances.[27] This, then, is the *quia* demonstration that the moon is a sphere. As to the *propter quid* explanation Grosseteste again is cryptic, stating merely that it follows from the moon's being "a homogeneous body," and this *per demonstrationem naturalem.* The reference to physical or natural demonstration is striking.[28] Does this mean that, for Grosseteste, physics can give a *propter quid* demonstration of an astronomical phenomenon? This is what the text says, and the interpretation coheres with others to be noted below; if so, however, Grosseteste's proofs clearly pertain to different sciences, and this is purportedly not what he has in mind. So perhaps *naturalem* should be read in the sense of a physical optics, again on the ontological suppositions outlined in *De luce,* where the *propter quid* explanation is that of a more metaphysical optics whereas the *quia* explanation is based merely on the visual geometry of the *perspectiva,* abstracting from its possible ontological implications.

Examples such as these are difficult to analyze because of Grosseteste's dual loyalty to Aristotelianism and to Neoplatonism, which systems provide quite different explanations of the phenomena he is considering. The cases involving two different sciences, on the other hand, present less of a problem because they pertain to what may be broadly called "mathematical physics,"[29] which makes use of mathematics as the *scientia subalternans* and of natural science as the *subalternata;* here the Aristotelian and Neoplatonic interpretations are more in harmony with each other. The two examples already mentioned are the rainbow and the healing of a wound, and to these we will further add Grosseteste's analysis of reflection in a mirror, as casting further light on his understanding of the complex relationships between sciences *propter quid* and *quia.*

Sciences related to each other as subalternating and subalternated can, according to Grosseteste, yield respectively knowledge *propter quid* and *quia* "about the same thing." [30] Examples he cites include geometry as related to geometrical optics and arithmetic as related to harmony; in these and similarly related disciplines the subject of the subordinated science, say geometrical optics, is made from the subject of the subordinating science, geometry, by "a super-added condition," in this case, that the lines and figures considered are not simply lines and figures but are rather those generated by visual rays.[31] The "same thing" about which these two sciences then demonstrate Grosseteste sees as equivalently made up of "two natures," the nature the thing receives from the subalternating science and the nature that is superadded to this. He is careful here to point out that the superior or subalternating science says nothing of the causes of the nature that is superadded, since this strictly pertains to the inferior or subalternated science; it treats only of the causes of its own nature.[32] Finally he notes that a science that is subalternated under one aspect may itself be subalternating under another aspect, and here he cites Aristotle's example of geometrical optics: while subalternated to geometry, it can subalternate to itself the meteorological science of the rainbow, and thus can be seen as intermediate between the two.[33]

The foregoing prepares for a brief analysis of the rainbow in the commentary, which is developed at much greater length in a special opusculum entitled *De iride.* Since this opusculum has been discussed in detail elsewhere,[34] it will be sufficient to note that for Grosseteste geometrical optics supplies *propter quid* knowledge of the rainbow, whereas physics (or meteorology) provides only *quia* knowledge.[35] The *propter quid* demonstrations, however, to be consistent with the foregoing, can relate only to the geometrical properties of the bow, which the meteorologist *qua* physicist knows only in *quia* fashion. Grosseteste does not raise the question whether the physicist himself can attain to causal, *propter quid* knowledge of the other (physical) properties of the bow that would be unknown to the geometrical optician, but again to be consistent with the foregoing he should allow this possibility.[36] If so, physics would not be limited a priori to *quia* demonstration, but could itself discover proper causes and, on this basis, would yield *propter quid* knowledge, though not of its subject's mathematical properties.

It is at this point that recourse to a previous example provided by Grosseteste in the commentary may prove helpful. The text is that immediately following the discussion of the demonstrability of eclipse *qua* corruptible and is concerned with the problem whether the middle term in a strict demonstration must be immediate, or appropriate, to the terms expressed as the subject and predicate of the conclusion.[37] Here Grosseteste discusses the defects in Bryson's so-called proof for the quadrature of the circle and then elaborates how, despite these, a strict demonstration along analogous lines may still be possible (though not of Bryson's conclusion), provided only that one premise pertains to a subalternating science and the other to a subalternated science.[38] What is required here, Grosseteste says, and this is completely consistent with what has been discussed above, is merely that the middle term be such that it incorporates elements properly pertaining to both sciences or, to use an alternate terminology, that it be equatable with the "two natures" already mentioned, the one the nature dealt with in the subalternating science and the other the nature that is superadded in the subalternated science.[39] Grosseteste's explanation of this is somewhat cryptic, and he offers to make it more evident by an example. Fortunately he chooses here to discuss the proof of a theorem in optics, namely, that the angles of incidence and reflection from a plane mirror are equal, and he treats the case rather thoroughly.[40] An analysis will therefore be helpful for clarifying his ideas on the role of mathematics in demonstrations pertaining to the physical sciences.

As it turns out, Grosseteste actually offers three proofs that are related to the theorem. The first two pertain to geometry and to geometrical optics respectively; in exposing these he uses the basic ideas from the *Catoptrica* attributed to Euclid,[41] gives a purely geometrical proof, then explains that this has to be modified and applied to visual or radiant lines, angles, and triangles, and finally offers a second proof that yields the desired result. The first demonstration, that of pure geometry, he says is *propter quid*, whereas the second, that of geometrical optics, is only *quia*; in other words, regarding the equality of the angles of incidence and reflection the optician knows only the fact, whereas the geometer knows the reasoned fact.[42] Up to this point, everything that Grosseteste has said about reflection is comparable with his analysis of the rainbow. Now, however, he goes on to raise the further question as to whether there can be a cause of the equality of the angles that is *not* a middle term taken from geometry but rather pertains to physics. His answer, surprisingly, is in the affirmative. Thus he states:

> The cause of the equality of the two angles made on the mirror by the incident ray and the reflected ray is not a middle term taken from geometry, but its cause is the nature of radiation generating itself along a rectilinear path, which, when it encounters an obstacle having in itself the nature of spiritual humidity, functions there as a principle regenerating itself along a path similar to that along which it was generated. For, since the operation of nature is finite and regular, the path of regeneration must be similar to that of generation, and so it is regenerated at an angle equal to the angle of incidence.[43]

This, as can be seen, is actually a third proof of the theorem and it is taken from physics or, if you will, from physical optics, in the sense that it explains an optical phenomenon in terms of its physical or natural causes.

So we see that the answer one might expect in the case of the rainbow, and that is contained implicitly in the opusculum *De iride,* is quite clearly stated in the discussion of reflection from a plane mirror. Rightly or wrongly, Grosseteste was apparently convinced that physical science, i.e., science only such in the generic sense, can arrive at true and appropriate causal explanations of physical phenomena, and in this sense can be a *propter quid* science. Precisely how it can do this, and how the process may be related to verification and falsification, requires further examination, but before proceeding to this we must mention the final example discussed by Grosseteste in the context of *propter quid* vs. *quia* demonstration, that namely of the healing of wounds.

This case, for Grosseteste, is quite different from all the others discussed.[44] He provides it as an illustration of the manner in which one science can know the causes of another science even though the two are not subalternated. The example is originally Aristotle's, who discusses the phenomenon of a circular wound's healing more slowly than wounds of other shapes, and notes that the physician knows only the fact (*quod*) whereas the geometer knows the reasoned fact (*propter quid*).[45] Grosseteste explains this as follows: the proper concern of the physician, viz., health, is not related in any way to magnitude, and thus cannot be studied by a geometer; yet "the shapes of wounds come under shapes simply speaking, of which geometry treats," and therefore "the causes of dispositions of wounds that are associated with their shape" can be known by the geometer.[46] Thus he concludes:

> The geometer knows that circular wounds are healed more slowly in knowing that of all perimeters the greatest is the circle and that the circle is the shape whose sides are farthest apart along every dimension, for it is on account of this that the sides of a circular wound go together with greater difficulty.[47]

Evidently it pertains to another theoretical science to discuss the proper (as opposed to the incidental) causes of health or healing, to which medicine would itself be subordinated as a science. And significantly, Grosseteste himself is aware of this, for he mentions earlier in this particular section of the commentary, and in this he is probably inspired by Avicenna, that such a science is the "science of the elements." He states:

> The science of the elements descends into the science of medicine, of which the subject is the human body as regards restoring it to health and removing health from it, yet not the human body as being itself an element but as being composed of elements. Therefore, with such sciences of which one is under the other, the superior science provides the reason (*propter quid*) for that thing of which the inferior science provides the fact (*quia*).[48]

Here, as in the analysis of reflection, Grosseteste is quite willing to admit that a physical science, that "of the elements," can uncover the causes of natural phenomena and thus be productive of *propter quid* knowledge.

Let us return at this point to the frequently stated account of Grosseteste's methodological contribution to the origins of modern science. According to this, physical explanations can generate at best *quia* knowledge, and even these are only probable, so that they require an additional process of verification and falsification, and this in turn leads to experimentation or at least to an experimental mentality. As we have seen, the first part of this interpretation is questionable, since for Grosseteste the natural philosopher can uncover true physical causes, as opposed to mathematical causes, and can therefore generate *propter quid* knowledge. If so, whence comes the need for verification and falsification?

The answer with respect to falsification is the easier of the two, since this was already part of the Aristotelian tradition and had been incorporated into the texts on which Grosseteste was commenting. Falsification is necessary, in classical logic, when strict demonstration is impossible of attainment and when the best that can be done is to propose alternative possible explanations; in such a situation some of the proposed explanations may perhaps be shown to be untrue because they do not agree with the facts, and thus they can be rejected. The procedure involved illustrates the mode of *tollendo tollens*, one of the legitimate ways of arguing with the hypothetical syllogism. Grosseteste, of course, shows himself familiar with this procedure and in fact uses it when discussing, in his *De cometis et causis ipsarum*, four theories about comets current in his time.[49] All of these he rejects, but only to substitute in their place an explanation that the modern scientist would find equally implausible. In this procedure he effectively has made no advance beyond the generally accepted Aristotelian methodology for dealing with hypothetical causal explanations.

With regard to verification the case is more interesting because Aristotle's way of dealing with the hypothetical syllogism would not allow such a procedure, since it would involve an illegitimate mode leading to the "fallacy of the consequent" (*fallacia consequentis*).[50] If Grosseteste can be shown to have worked out, either explicitly or implicitly, a procedure for verifying theories, particularly through experimental test, one would have to admit a considerable advance beyond Aristotelian methodology. Yet here a study of Grosseteste's writings proves disappointing: he has no explicit discussion of verification as a procedure,[51] and the instances of verification that have been alleged in his works by various historians of science prove, on examination, to involve little more than Aristotle's general insistence that theories should be accepted only if what they affirm agrees with the observed facts.[52]

One of the better examples of so-called verification to be found in Grosseteste is the discussion in his commentary on the *Posterior Analytics* of the purgation of red bile by scammony, a plant native to Asia Minor.[53] Here Grosseteste points out that an observer will first notice only the coincidence of the eating of scammony with the discharge of red bile; then constant observation will lead him to formulate a *universale experimentale* wherein he sees a causal connection between scammony's ingestion and the purgation that follows. Finally, by isolating and excluding all other likely causes and by repeatedly administering scammony alone he will become convinced that "all scammony of its nature withdraws red bile." [54] This is an interesting example, since it connotes a type of controlled experimentation characteristic of modern scientific method, and one is tempted to ask if Grosseteste actually experimented in this fashion. The answer turns out to be that this is highly unlikely, since the phenomenon was already recorded by Galen and since Avicenna has a rather full discussion of this methodological point and instances the same example in at least two places in his writings.[55]

What this suggests, moreover, is that Grosseteste's treatment of empirical method was quite speculative, as one might suspect, and was not applied carefully in his own scientific investigations. This explains why his only quantitative law in optics, namely, the law of refraction wherein he states that, when a light ray passes from a less dense to a more dense medium the angle of refraction will be one half the angle of incidence,[56] is in error. Crombie shows disappointment at this, since "very simple experiments could have shown Grosseteste that his quantitative law of refraction was not correct." [57] The mistake here consists in thinking that Grosseteste was committed to experimentation in a systematic way.[58]

As we have seen, his "metaphysics of light" and its incipient rationalism made him quite confident of the mind's ability to discern the physical causes of optical phenomena, and thus he did not feel any particular need to have recourse to experiment. Like many medievals, Grosseteste used the terms *experimentum* and *experientia*, but these terms did not have for him the connotations they have today. *Experimentum*, in fact, is well translated by the modern English word, "experience," and it frequently represents a quite eclectic selection of examples, some based on the author's own experience and others based on reports either heard or read, but usually given equal credence regardless of source.[59] Thus one must be wary of seeing in examples like that of scammony and red bile an adumbration of controlled experimentation as used to verify theories in contemporary science.

What, then, are we to make of Grosseteste's commentary on the *Posterior Analytics* and its relationship to modern scientific method? This work undoubtedly stimulated speculation about the causes of physical phenomena, and if in fact it was defective in its explanations, from that very circumstance gave even greater stimulation to finding out what the correct explanation might be. Grosseteste read Aristotle, as we have noted, with a Neoplatonist bias; even apart from that, however, he was not working with the best translation, and this itself led to questionable interpretations. A final example, of interest for the detail with which it analyzes the phenomena of thunder and lightning, may serve to make this point. It will also be of value for showing how Grosseteste understood *propter quid* demonstration to involve all four causes, and particularly how the four causes are related to each other in the demonstrative process.

Like Aristotle, Grosseteste explains the relationship of particular causes to demonstration when comparing causal definitions and explaining how such definitions may be related to one another through the process of demonstration.[60] In this context he explains that a definition based on the formal cause may be used to demonstrate a definition based on the material cause, and that if the definition contains both the formal and the material causes it is equivalent to a demonstration and differs only in the way in which the terms are arranged. He also allows the possibility of definitions and demonstrations through efficient and final causality, and favors including the final cause in the formal definition and the efficient cause in the material definition.[61] There are some Neoplatonic overtones in this exposition, but otherwise it is a straightforward commentary on Aristotle. Grosseteste encounters a problem, however, when he comes to Aristotle's example, which involves the causes of thunder and how these are related to formal and material definition and to the demonstrative process.

To understand the problem it is necessary to know that Aristotle, in his *Meteorology*, teaches that thunder is caused by the forcible ejection of dry vapors trapped in clouds and being rapidly compressed; the dry vapors, on being ejected, usually catch fire, and this produces lightning.[62] Although the lightning flash is seen after the thunder is heard, this is not the order in which the process occurs: the sound comes first, then the lightning, and our observation of the reverse sequence is accounted for by the slow velocity of sound with respect to light, which is transmitted instantaneously. Grosseteste gives evidence of knowing this explanation, and thus is somewhat surprised when he reads the example given by Aristotle in the *Posterior Analytics*, namely, that the cause (*propter quid*) of thunder is the quenching of fire in a cloud, whereas thunder itself is the rumbling sound in the cloud.[63] Here it appears that, rather than have thunder the cause of lightning, as stated in the *Meteorology*, Aristotle makes lightning the cause of thunder. "The order of Aristotle's words is here mixed up," writes Grosseteste, and in the light of his rearrangement he continues, "but I believe that they should be rearranged as I have done so that his intention will become clear." [64] The instance is typical of a medieval commentator's encountering what to him is a textual difficulty, but which will become the occasion for his delving deeper into the matter at hand in order to obtain a better explanation.

Grosseteste's solution is that "rumbling sound in cloud" is the formal cause of thunder, whereas "quenching of fire in cloud" is the material cause. Since the formal cause can be used to demonstrate the material cause, one can say that the "quenching of fire in cloud" is "on account of" (*propter*) the "rumbling sound in cloud." His justification of this is as follows:

> I say that the form is the end of the matter and that matter in truth does not effect form but functions as an occasion for the efficient cause to act and induce the form in the matter. Hence the expression, "on account of" [*propter*], as it is understood in a composed definition when the material and formal definitions are taken together, does not indicate the operation of efficient causality but rather that of final causality. For universal nature quenches fire in cloud so that a rumbling sound may arise in cloud.[65]

Grosseteste then goes on to detail the operation of the various causes in the production of thunder, to show that the "efficient cause of the noise is not the quenching of fire, but the latter is a necessity on the part of the matter, without which rumbling would not occur." [66] From his detailed discussion the following emerge as the four causes of thunder. [1] The formal cause is "rumbling sound in cloud," which is *per se nota* and cannot be demonstrated. [2] The material cause, briefly, is "quenching of fire in cloud," although this is closely associated with the operation of efficient causes and must be understood as their necessary material condition. [3] The efficiency involved is described through the mechanism that produces the thunder. The influence of the heavenly bodies causes the cold wet vapor that is the matter of clouds and rains to ascend, and with it also ascends the cold dry vapor that is the material of wind; both of these vapors come to an intermediate cold region of air that compresses the vapors; there is also considerable fire in the two surrounding layers of air. The compressing cold vapors solidify suddenly in the cloud, and compress dry vapors that they trap in the middle; under this compression the dry vapors explode, rupture the cloud and escape, and in so doing they rarefy and ignite. The fire is extinguished by the vapor that remains in the cloud, but this heats the vapor and thins it out, and as it hisses through the heavier surrounding vapors it makes a rumbling noise. The noise itself is the vibration of the parts of the escaping gas, and it is caused by a natural power resident in the gas as the parts ejected by compression return to their natural place. The efficient cause of the noise, therefore, is not the quenching of fire, since this is only a material necessity requisite for the noise's production. Rather the efficiency involved is traceable to the natural capabilities of gases in such circumstances and ultimately to the influence of the heavenly bodies.[67] [4] According to the Pythagoreans, the final cause of thunder is to terrify those being detained in the infernal regions. Grosseteste is non-committal on this, but he himself suggests an alternate explanation: perhaps nature makes it thunder to purify the air and separate out extraneous qualities, because with the generation of thunder there is an intense vibration of similar parts that crunches them together and forces out all impurities.[68]

The foregoing identification of the four causes of thunder goes considerably beyond the text of the *Posterior Analytics*.[69] Grosseteste actually uses it as an occasion to refine and correct the peripatetic explanation in the *Meteorology*, while attempting to remain faithful to Aristotle's methodological canons. And at least part of his reason for doing so seems to be that he is not sure of the text on which he is commenting — the actual ordering of Aristotle's words may have gotten reversed. Interestingly enough, commenting on the same passage some forty or fifty years later and probably with the aid of William of Moerbeke's corrected version, Thomas Aquinas sees no difficulty. He recognizes that the ordering is the reverse of what one would expect from Aristotle's teaching, but correctly identifies it as the opinion of Anaxagoras and Empedocles and explains the discrepancy by noting simply that Aristotle "frequently uses the opinions of others in his examples." [70]

Aquinas, of course, had more to work with than did Grosseteste, whose attempts to understand the *Posterior Analytics* were pioneering. Grosseteste thought that he had a privileged insight in terms of his "metaphysics of light," which would yield *propter quid* knowledge not only of the geometrical aspects but also of the physical aspects of many natural phenomena. But when he descended into the complex details of such common phenomena as thunder and lightning, he could not supply the convincing explanations that would parallel those of geometrical optics. Perhaps the very inadequacy of his examples stimulated his students to continue their search for causes, and to use empirical methods in so doing. If so, they would still remain under the dual influence of Neoplatonic geometrism and Aristotelian experimentalism, understanding the latter as the search for generalizations based on one's own experience or that recorded by others.

a. Roger Bacon

This methodological influence of Grosseteste is best seen in the work of two of his followers, Roger Bacon [71] and John Peckham.[72] Of the two, Bacon has received the greater notoriety and was the more influential protagonist of Grosseteste's teachings. This is not to say that Bacon was a thorough-going disciple of the Bishop of Lincoln; critic that he was, he eventually turned against this man from whom he claimed to have learned so much and disagreed on details of his teaching. But in the main, Bacon emulated Grosseteste in his love for languages and his desire for accurate translations, in the great importance he placed on mathematical reasoning for probing the secrets of nature, and finally, to a degree not conspicuously evident in his preceptor, in his insistence on experience and experimental science as necessary complements to mathematical ways of reasoning.

Part IV of Bacon's *Opus maius* is concerned with mathematics, and here he is intent to show that this discipline is not only essential for understanding all other sciences, but that without it one cannot understand the things of this world. He writes: "For the things of this world cannot be known without a knowledge of mathematics. For this is an assured fact in regard to celestial things, since two important sciences of mathematics treats of them, namely theoretical astronomy and practical astronomy. . . ." [73] He then goes on to explain how mathematics is essential to these astronomical disciplines. Following this, he discourses on the way in which mathematics is of equal importance for learning of terrestrial things, and here his reasoning is important because it links mathematical reasoning with the Aristotelian doctrine of causes. Bacon explains:

> It is plain, therefore, that celestial things are known by means of mathematics, and that a way is prepared by it to things that are lower. That, moreover, these terrestrial things cannot be learned without mathematics is clear

from the fact that we know things only through causes, if knowledge is to be properly acquired, as Aristotle says. For celestial things are the causes of terrestrial. Therefore these terrestrial things will not be known without a knowledge of the celestial, and the latter cannot be known without mathematics. Therefore a knowledge of these terrestrial things must depend on the same science. In the second place, we can see from their properties that no one of these lower or higher things can be known without the power of mathematics. For everything in nature is brought into being by an efficient cause, and the material on which it works, for these two are concurrent at first. For the active cause by its own force moves and alters the matter, so that it becomes a thing. But the efficacy of the efficient cause and of the material cannot be known without the great power of mathematics, even as the effects produced cannot be known without it. There are then these three, the efficient cause, the matter, and the effect. In celestial things there is a reciprocal influence of forces, as of light and other agents, and a change takes place in them, without however any tendency toward their destruction. And so it can be shown that nothing within the range of things can be known without the power of geometry. We learn from this line of reasoning that in like manner the other parts of mathematics are necessary; and they are so for the same reason that holds in the case of geometry; and without doubt they are far more necessary, because they are nobler. If, therefore, the proposition can be demonstrated in the case of geometry, it is not necessary . . . that mention be made of the other parts.[74]

Following this general statement, Bacon discourses in detail on how the efficient cause produces effects in various subject matters by the "multiplication of species," and how it does this acting along linear paths that are particularly suited to geometrical demonstration.[75]

Since Bacon, like Grosseteste, thus subscribes to a "metaphysics of light," it is not surprising that Part V of the *Opus maius* is devoted to optical science. What is remarkable in this part of the *Opus maius* is the great detail with which Bacon analyzes the examples given by Grosseteste in his commentary on the *Posterior Analytics,* several of which have already been mentioned. For example, he discusses the optics involved when objects are seen at great distances and the various illusions that can be involved.[76] He has a quite detailed analysis of twinkling, or scintillation, and implicitly corrects Grosseteste on several points, going considerably beyond him in explaining the way in which the object seen and the medium through which the visual rays are transmitted can affect the scintillation.[77] In this discussion he shows himself well acquainted with Averroës's views, which it appears were not known to Grosseteste at the time of his writing. Bacon acknowledges having superior translations of Aristotle, and in the context of the discussion of the planets' nearness and the relation of this to their non-twinkling, he states with confidence, "This is the opinion of Aristotle, as we clearly infer from several translations, and especially from that purer one that has been made directly from the Greek." [78] Another example is the very detailed explanation that Bacon gives of the phases of the moon in terms of the visual cone and the same general optics as had been briefly sketched by Grosseteste.[79] Again, his analysis of reflection and refraction covers the same ground as Grosseteste's treatment of these phenomena, but with greater detail and a noticeable development of thought.[80] The same may be said for Bacon's analysis of the rainbow, to which we shall return shortly.

Bacon's commitment to mathematics, as can be seen from the foregoing, was not significantly different from Grosseteste's. With regard to experimental science, however, Bacon shows a more pronounced emphasis, although it is questionable whether this extends to his own practice or is merely a commitment in principle. Part VI of the *Opus maius* is devoted precisely to the subject of experimental science, and here Bacon is quite insistent that mathematical reasoning is not sufficient of itself, but must be supplemented by an appeal to experience in order to make its intuitions certain. As he states it: "Reasoning draws a conclusion and makes us grant the conclusion, but does not make the conclusion certain, nor does it remove doubt so that the mind may rest on the intuition of truth, unless the mind discovers it by the path of experience . . ." [81] In this context he gives his much-quoted example:

For if a man who has never seen fire should prove by adequate reasoning that fire burns and injures things and destroys them, his mind would not be satisfied thereby, nor would he avoid fire, until he placed his hand or some combustible substance in the fire, so that he might prove by experience that which reasoning taught. But when he has had actual experience of combustion his mind is made certain and rests in the full light of truth. Therefore reasoning does not suffice, but experience does.[82]

Having said this, Bacon immediately extends his example to mathematical reasoning, stating that "the mind of one who has the most convincing proof in regard to the equilateral triangle will never cleave to the conclusion without experience . . ." [83] And with regard to the application of mathematics to astronomical sciences, he makes the further statement:

Mathematical science can easily produce the spherical astrolabe, on which all astronomical phenomena necessary for man may be described, according to precise longitudes and latitudes . . . but the subject is not fully explained by that device, for more work is necessary. For that this body so made should move naturally with the daily motion is not within the power of mathematical science. But the trained experimenter can consider the ways of this motion, aroused to consider them by many things which follow the celestial motion, as, for example, the three elements which rotate circularly through the celestial influence . . . ; so also comets, the seas and flowing streams, marrows and brains and the substances composing diseases. Plants also in their parts open and close in accordance with the motion of the sun. And many like things are found which, according to the local motion of the whole or the parts, are moved by the motion of the sun. The scientist, therefore, is aroused by the consideration of things of this kind, a consideration similar in import to that in which he is interested, in order that at length he may arrive at his goal.[84]

This statement has a far more empirical cast that anything encountered in Grosseteste, and yet from the examples given and from Bacon's actual contribution to the advance of science, one can well wonder to what extent Bacon was committed to experimentation as an actual scientific practice.[85]

b. John Peckham

John Peckham may have been influenced by Bacon, and he followed the general orientation of Grosseteste's school, although he devoted much of his effort to theology and in so doing opposed the rising tide of Aristotelianism with his own more orthodox Augustinianism.[86] His *Perspectiva communis*, however, is a very systematic working out of the principles of geometrical optics, so much so that it was used as a textbook in the universities up until the sixteenth century. The work is greatly indebted to Alhazen's *Perspectiva*, and in fact may be regarded as a compendium of Alhazen's much longer treatise. Peckham expresses his purpose in writing the *Perspectiva communis* as follows:

> I shall compress into concise summaries the teachings of perspective, which [in existing treatises] are presented with great obscurity, combining natural and mathematical demonstrations according to the type of subject matter, sometimes inferring effects from causes and sometimes causes from effects, and adding some matters that do not belong to perspective, although deduced from its teachings.[87]

As can be seen from this, Peckham follows the Grossetestian tradition of giving both physical and mathematical demonstrations, and quite clearly analyzes optical phenomena in terms of their causes and effects, while not limiting himself exclusively to the a priori or to the a posteriori approach. There can be no doubt, however, that in so doing he searches for the causes of all visual phenomena, and tries to ground his reasoning as best he can in observation and in effects that can be directly verified. Many of the topics discussed by Grosseteste and Bacon, such as the twinkling of stars, the phases of the moon, and the properties of the rainbow, are given more nuanced and fuller treatment by Peckham in the context of this systematic development.[88]

There can therefore be little doubt that both Bacon and Peckham progressed beyond Grosseteste, particularly in their understanding of optical phenomena. Some of this work, it would appear, required either experimentation or a judicious selection of factual information reported by others. The extent of these advances are best seen in the discussions by Bacon and Peckham of the structure of the eye and the refraction and reflection of light rays in spherical lenses and mirrors. Lindberg has shown that Bacon's work on the rainbow, regarded by some as a regression from Grosseteste's contribution, actually constituted an advance in the role it assigned to the individual raindrop.[89] Both Bacon and Peckham delved deeper into the causes of refraction, supplying explanations that show similarities with the mechanistic theories of later centuries.[90] But it was in the theories of pin-hole images that both Bacon and Peckham showed the most originality, and this coupled with a recourse to observed facts, although at least one of Bacon's so-called facts is erroneous and could only have been based on an imaginary experiment or a construction deduced from one of his diagrams.[91]

The Oxford scholars of the thirteenth century thus made important contributions to optical sciences, and in so doing stressed the role of mathematics and experimentation in their causal analyses. The "metaphysics of light" provided the basic framework for their reasoning, and under this influence a considerable amount of rationalism and mathematicism is manifest in their methodological procedures. As far as experimentation is concerned, there seems to be no evidence that Grosseteste himself felt the need for this. Bacon and Peckham, on the other hand, were more empirical in their approach, although their appeal seems to have been more to experience than to anything that might be called experimentation in the modern sense.[92] Yet all three were committed to a realist philosophy of science; they were truly attempting to find causes for the effects they observed, and they had little or no doubt that such causes were the ontological antecedents and the proper explanations of the phenomena they sought to explain. The fourteenth century at Oxford, as we shall now see, questioned this open acceptance of realism and inclined instead to the nominalism of William of Ockham. The mathematical approach to reality was able to survive this change, and indeed benefited from it in some ways, but the experimental component was not similarly favored and, if anything, fell even further into disuse than in Grosseteste's work at the onset of the thirteenth century.

2. *The Mertonians*

The group of writers associated with Merton College from the 1320's to the 1360's at Oxford are more frequently cited as precursors of modern science than Grosseteste and his followers. These men were heirs, of course, to the Grossetestian tradition and thus were no strangers to applications of mathematics in philosophical and even theological reasoning. At the same time, they came under the influence of a revolutionary type of thinking, that associated with the nominalism of William of Ockham, which purported to give a more authentic interpretation of Aristotle than those current in the thirteenth century. The resulting mixture of Aristotelian and Ockhamist thought enabled them to develop mathematical analyses remarkably similar to those employed in seventeenth-century science. Through their use they advanced considerably beyond the geometrical optics that had been grounded in the "metaphysics of light" and turned attention instead to difficult problems of local motion, particularly those later identified as pertaining to kinematics and dynamics. The resulting development can conveniently be traced by sketching first the essential features of William of Ockham's contribution and the disagreement between him and Walter Burley, and then by summarizing the thought of those more proximately concerned with the evolving science of mechanics, namely, Thomas Bradwardine, William Heytesbury, John Dumbleton, and Richard Swineshead.[93]

a. William of Ockham

William of Ockham's ideal of science was clearly that of the *Posterior Analytics* of Aristotle, but he coupled this with distinctively new philosophical attitudes toward the role of causality in the physical universe and man's ability to comprehend the same.[94] Being also a theologian with a strong faith commitment, Ockham placed great emphasis on the absolute power of God and on the fact that the physical uni-

verse is completely dependent on God's will. His underlying
voluntarism led him to hold that there could be no inherent
necessity determining anything in the universe to be exactly
as it is. He held that all things, excepting only those that involve contradiction, are possible for God, and that whatever
God produces by secondary causes He can also produce and
conserve directly without their causality. Another of his distinctive teachings was that God can cause and conserve any
entity or reality, whether this be substance or accident, apart
from any other entity or reality.[95] Unlike the scholastics who
preceded him, however, Ockham would not ascribe reality to
"common natures" or to all the Aristotelian categories, re-
stricting this ascription to individual "absolute things" (*res
absolutae*), which in his view could be only substances and
qualities. "Besides absolute things, namely, substances and
qualities, nothing is imaginable, neither actually nor
potentially." [96] The remaining categories used by philosophers to discuss natural things were regarded by Ockham as
abstract nouns, used for the sake of brevity in discourse. As a
consequence, such words as quantity, motion, time, and
place, even velocity and causality, had for him no real refer-
ent over and above an individual substance or its qualities.

One consequence of this view is that quantity, for Ockham, became merely a term denoting either a substance or a
quality, granted with the connotation that the substance or
quality in question has parts that are distant from each other.
From the point of view of real existence, "there is no quantity
different from substance or quality, just as there is no thing
with part locally distant from part except a substance or a
quality." [97] Influenced by this view of quantity, it is perhaps
not surprising that those who subscribed to Ockhamist doctrines could develop mathematics and calculatory techniques
without seeing the quantities with which they were concerned as having direct counterparts in the real world. Quantification became for them more a matter of logic and language than of physical science; perhaps for this reason,
Ockham's followers soon became involved in all manner of
sophismata calculatoria, which dwelt interminably on logical
subtleties but had little value for promoting an understanding
of nature.

Another case in point is Ockham's conception of motion and the causality involved in its production. For him, as we have seen, motion cannot be an absolute entity distinct from the body in motion. Thus, to explain local motion all one need do is have recourse to the moving body and to its successive positions: "the phenomenon can be saved by the fact a body is in distinct places successively, and not at rest in any." [98] And if motion itself is only a relative entity, the same will have to be said for motor causality. This view enabled Ockham to offer a very simple solution to the projectile problem, which had embarrassed all Aristotelians up to his day. He rejected not only Aristotle's explanation but also that of the theory of impetus, which would assign a motive power to some quality implanted in the projectile by the thrower. In his view the motor causality principle, *Omne quod movetur ab alio movetur,* is not applicable to local motion, since local motion is really not a new effect. He writes:

> A local motion is not a new effect, neither an absolute nor a relative one. I maintain this because I deny that position (*ubi*) is something. For motion is nothing more than this: the movable body coexists with different parts of space, so that it does not coexist with any single one. . . . Therefore, though every part of the space which the movable body traverses is new in regard to the moving body — insofar as the moving body is now traversing these parts and previously it was not — nevertheless no part is new without qualification.[99]

Some authors have seen in this type of reasoning an adumbration of the principle of inertia, and rather strangely, in fact, since Ockham had no intention of limiting himself to uniform motion in a straight line.[100] As he saw it, there was no need to seek the cause of *any* motion, whether uniform or accelerated, since it was not the type of entity to require a cause.

b. Walter Burley

Being a Franciscan friar, Ockham could not have been a member of Merton College, which was restricted to seculars, although his ideas were very influential among the Mertonians.[101] One of the earliest fellows of Merton College to take an interest in Ockham's view, and this mainly to refute them, was Walter Burley. Burley was older than Ockham, and his philosophical and theological views were more traditional, although he too worked extensively in logic and, according to historians of that discipline, made significant contributions to its development.[102] Burley commented on all of the logical works of Aristotle, as well as on the *Isagoge* of Porphyry and the *Liber de sex principiis* attributed to Gilbert de la Porrée; his work includes a long commentary on the *Posterior Analytics*, a series of questions on the same, and an abbreviation of this teaching, which may or may not have been written by him but is printed with the commentary of Grosseteste in the Venice edition of 1514. Also of interest are Burley's commentaries and questions on the *Physics* of Aristotle and on his remaining *libri naturales*. These show that Burley was an assiduous student of Grosseteste and of Albertus Magnus, among others, thus preserving the thirteenth-century tradition while at the same time opening up new problems in natural philosophy that were susceptible to logical analysis in a more realist framework than that provided by Ockham.

Two topics relevant to the foregoing discussion are Burley's quodlibetal question, *De primo et ultimo instanti*, and his views on the nature of motion particularly as detailed in his two treatises on the intension and remission of forms.[103] *De primo et ultimo instanti* is concerned with the problem of designating the first and last moments of a motion or change that takes place within a given time interval. Burley's solution is worked out within an Aristotelian framework which would allow that the motion be limited either intrinsically at beginning and end (*incipit et desinit*) by an instant that belongs to the time interval corresponding to the motion, or extrinsically by a first or last instant that does not pertain to this interval. Granted this, and making precisions regarding the mutual exclusiveness of intrinsic and extrinsic limits, Burley attempts to decide just what kinds of things are characterized by the different types of limiting instants. In so doing he makes the distinction between a successive entity (*res successiva*) and a permanent entity (*res permanens*). Motion itself is a *res successiva*, and for Burley, as for Aristotle, this can only be terminated by extrinsic instants at its beginning and at its end. There are many complexities to his doctrine, particularly as it relates to *res permanentes* undergoing change, and this specifically in the context of his own special theory of the intension and remission of the forms that may be involved, but these need not concern us here.[104]

Suffice it to note that Burley's distinction between *res permanentes* and *res successivae* is quite different from that between Ockham's *res absolutae* and *res relativae*. Whereas motion, for Ockham, was only a *res relativa*, and as such was really nothing more than the object undergoing change and the place, quality, or quantity that it may successively acquire, Burley maintained that motion was a *res successiva* and as such was something real over and above the connotations associated by Ockham with the moving object. This difference between Burley and Ockham may be explained with the aid of the much used fourteenth-century distinction between motion viewed as a *forma fluens* and motion viewed as a *fluxus formae*.[105] Since local motion, for Ockham, was nothing more than the new positions or forms of *ubi* successively acquired by the moving object, the motion itself could be viewed as a *forma fluens*. The form or terminus acquired as a result of the motion, in Ockham's view, somehow included all of the forms successively traversed along the way. Burley allowed the validity of this analysis, but contested its completeness, maintaining that motion is something over and above the form or terminus acquired, being itself the very flux or transmutation of the subject whereby the form or terminus comes to it. Thus, for him, motion is not only a *forma fluens* but also a *fluxus formae*. This allowed him to maintain that every instant of the motion corresponds to a different form, that there is no form, strictly speaking, that corresponds to all of them, and that the successive forms are neither parts of, nor contained within, the form that terminates the motion.[106]

One important consequence of this difference in viewpoints was that, whereas Ockham had effectively denied the reality of the motion and any causality involved in its production, Burley could maintain that local motion is an *ens reale* and as such has its own proper causes and effects. This difference of viewpoint was stereotyped by referring to Ockham's conception as the *via nominalium* and to Burley's as the *via realium,* a difference that was to be vigorously debated throughout the fourteenth and subsequent centuries. Needless to say, it became crucial for all subsequent discussions of causality, particularly as related to the science of motion.

c. Thomas Bradwardine

When one speaks of the fourteenth-century Mertonians and their influence on the science of mechanics, he usually has in mind Thomas Bradwardine and other fellows of Merton who worked in the context provided by Bradwardine's *Tractatus de proportionibus velocitatum in motibus,* namely, Heytesbury, Dumbleton, and Swineshead. Bradwardine himself was an accomplished philosopher and theologian, later serving as Archbishop of Canterbury, but his great contribution consisted in applying mathematical analysis to physical problems in a way that would have been applauded by Grosseteste and Bacon, though going far beyond their rudimentary efforts.[107] It would seem that Bradwardine was not too sympathetic to Ockham, arguing particularly against his theological views, but also concerning himself with a question based on Aristotle's *Physics* that for Ockham would have been meaningless.[108] Bradwardine's treatise was in fact concerned "with the ratios of speeds in motions" when such motions are viewed in relation to the various causes, i.e., the forces and resistances, that produce them. Thus the problem to which the entire treatise is addressed is how to correctly relate a variation in the speeds of a moving object, which is what the "ratio of speeds" in the title indicates, to variations in the forces and resistances that determine these speeds. How Bradwardine worked out his answer need not concern us here, although it may be mentioned that he equivalently proposed a geometrical proportionality between speed, force, and resistance that would be expressed in modern notation as an exponential or logarithmic function. The sophistication of his mathematical analysis in this treatise and in his *Tractatus de continuo* is truly remarkable, and undoubtedly accounts for its stimulating many disciples to work out further mathematical details relating to his theory of motion.[109]

Bradwardine's mathematical advance, however, was

made at the expense of his philosophical consistency in understanding motion, particularly the reality associated with it as a process. As Weisheipl has argued, effectively Bradwardine equated motion with "speed of motion," which he conceived as a qualitative ratio that could be intensified and remitted like any other quality, provided only that his law of geometrical proportionality would be verified.[110] Aristotle and thirteenth-century scholastics had viewed speed as a modality or property of motion, but not as motion itself, nor would they allow speed to be taken in such a univocal sense as to permit a direct mathematical comparison of motions of various types. Yet, whatever its philosophical difficulties, Bradwardine's identification of motion with speed was doubtless a necessary step in the mathematization of motion, and thus for the birth of a mathematical physics in the modern sense.

d. Dumbleton and Heytesbury

Although presented as a solution to a realist problem, Bradwardine's work was also open to nominalist interpretation; thus it is not surprising that two writers who came under his inspiration, John Dumbleton and William Heytesbury, were more or less Ockhamist sympathizers. Dumbleton's *Summa logicae et philosophiae naturalis* is one of the few systematic expositions of fourteenth-century physics, and it shows a curious mixture of Aristotle, Ockham, and Bradwardine.[111] Its author was obviously unaware of the inconsistencies, not to say contradictions, implied in the various views of motion to which he subscribed. Still his work performed a useful function in providing a more general interpretation of Bradwardine's law and by translating his relationships into the then-current language of the "latitude of forms," wherein all changes, qualitative as well as quantitative, were viewed as traversing a distance or "latitude" that would be readily quantifiable. According to Dumbleton's analysis, equal latitudes of motion or speed would always correspond to equal latitudes of the ratio of force to resistance involved in its production. Of all the Mertonians, it has been said, "Dumbleton was closest to Grosseteste's tradition of science, especially his interest in light."[112] He also employed "Grosseteste's method of resolution and composition, and the distinction between formal and material definitions." [113]

William Heytesbury likewise combined interest in physical problems with great proficiency in logic, and although not an outspoken defender of Ockham, nonetheless subscribed to the basic nominalist positions. This is particularly true of his views on the reality of place, time, and motion. As Wilson correctly observes, "for Heytesbury, the real physical world consists only of objects; point, line, surface, instant, time, and motion are *conceptus mentis*. These affirmations (or perhaps they are better termed 'negations') are in accord with the nominalist or terminist position, developed at length in William of Ockham's work on the *Physica*." [114] Heytesbury was particularly concerned with the *sophismata* that arose from attempts to denominate and calculate aspects of form associated either with its latitude or its intension and remission. It is in such a context that this Mertonian, conceiving motion after the fashion of a quality whose latitude or mean degree could be calculated, formulated one of the most important kinematical rules of fourteenth-century mechanics, since come to be known as the Merton "mean-speed theorem." The theorem states that a uniformly accelerated motion is equivalent to a uniform motion whose speed is equal throughout to the instantaneous speed of the uniformly accelerating body at the middle instant of the period of its acceleration.[115] The notion of instantaneous speed (*velocitas instantanea*) contained herein is of critical importance to the development of modern mechanics, and represents a significant departure from earlier methods of designating the speed of motion among scholastic thinkers.

e. Richard Swineshead

The most complete mathematical treatment of kinematics, however, is that to be found in the *Liber calculationum* of Richard Swineshead, who was so proficient in his subject that he was antonomastically referred to as "the Calculator" in subsequent literature.[116] Swineshead's work prescinds from philosophical disputes and considers in mathematical detail a wide variety of topics: problems relating to the intension and remission of qualities; intensification where the variation is non-uniform; problems of intensification relating to elements and compounds; rarefaction and densification; ways of denominating the speed of augmentation; action and reaction; problems of maxima and minima; details of how illumination takes place in various media and assuming light sources of varying intensity; and series of rules applicable to local motion affected by all types of forces and resistances, both uniform and difform. In most of this Swineshead is merely unfolding all of the implications of Bradwardine's rule when this is seen as capable of relating a wide variety of velocities, forces, and resistances. One interesting application of the rule, for example, is Swineshead's use of it to resolve a puzzle relating to the fall of a heavy body as it approaches the center of the universe.[117] If such a heavy body be conceived of as a thin rod (*corpus columnare*), the problem is whether its center would ever coincide with that of the universe, in view of the fact that as soon as any part of the rod had passed through the center, it would tend back toward the center and thus would constitute a resistance against the rod's continued motion. On the assumption that such a resistance does result from "parts beyond the center," Swineshead gives a thorough mathematical proof that, under the action of the forces in the light of Bradwardine's rule, the center of the rod can never coincide with the center of the universe. The proof is quite elegant, since it shows that the body can come arbitrarily close to, though never reach, the world's center in a finite time, and has rightly been described as a mathematical *tour de force*.[118] Yet, from this null result, Swineshead concludes only that a body, far from being simply the mathematical sum of its parts, must be regarded as an integral whole all of whose parts assist the whole in attaining its proper end. Small wonder that the editors and translators of this tractate conclude their exposition with the wry observation that it "ends with the frustrating spectacle of an author using sophisticated techniques of applied mathematics in order to show that in the problem at issue mathematics is inapplicable." [119]

Unfortunately there is little in the *Liber calculationum* relating to the entitative status of motion, or of its various causes. There is, however, an anonymous *Tractatus de motu*

locali difformi, thought by some to be the work of Richard
Swineshead, that gives a summary statement of the four
causes of local motion.[120] The translation reads as follows:

> For anyone wishing to understand motion, it is expedient
> to investigate its principal causes, which are the mate-
> rial, formal, efficient, and final. Hence the material cause
> of motion, or the matter of motion, is whatever is ac-
> quired through motion; the formal cause is a certain
> transmutation conjoined with time; the efficient cause is
> a ratio of greater inequality of the moving power over re-
> sistance; and the final cause is the goal intended.[121]

Here we have the traditional Aristotelian doctrine of the four
causes applied to local motion, but stated in terms of Bradwardine's ratios. As Weisheipl has observed, "this conception
of motion as a real ratio of distance to time, and produced by
a determined ratio of mover to resistance, is vastly different
from the nominalist interpretation offered by Ockham." [122] It
is typical of the Merton School, however, and shows how a
certain ambivalence had come to be accepted among these
fourteenth-century mechanicians, who could speak of motion as
having causes, and of causes themselves as being ratios, seemingly without realizing how much they were departing, in so
doing, from the basic philosophies of both Ockham and Aristotle.

3. *The Oxford Contribution*

From the viewpoint of the mathematical component of modern science, as represented by optics and mechanics, it is not
difficult to trace a certain continuity between the Grosseteste
group and the Mertonians. Both were concerned with explanations in mathematical terms, and both related these explanations in one way or another to Aristotelian causality. Yet
neither group was purely Aristotelian in its approach, since,
as we have seen, Grosseteste placed great reliance on an underlying Neoplatonic "metaphysics of light" when attempting
to supply ultimate justifications, whereas the Mertonians were
influenced by Ockhamist views that undermined their commitment to realism, particularly as relating to motion and its
causality.

The effect of both these departures from Aristotle was that the science developed at Oxford, the claims of some historians notwithstanding, was deficient in its attitude toward experimentation and empirical findings generally. As we have argued, Grosseteste's ontological commitments led him to a somewhat facile acceptance of *propter quid* demonstrations in optics, so that he never had to be concerned with the measuring techniques known to Alhazen nor with the experimental procedures later to be developed by Theodoric of Freiberg. Roger Bacon, it is true, combined an interest in Grosseteste's mathematicism with a strong advocacy of experimental method, but so far as is known he contributed little to the development of such method in the science of his day. John Peckham's systematization of optical science represents the furthest that Grosseteste's school could go, and this was essentially by way of providing causal and geometrical explanations for a good many optical phenomena accessible to ordinary human experience. This is not to deny that some elements of Aristotle's empirical approach survived in these thirteenth-century Oxonians. The point is rather that empiricism was not the hallmark of their approach to optical science. Their contribution was essentially mathematical, in that they showed how the *Posterior Analytics* could be made to work in terms of a geometrical optics, and thus stimulated the search for similar causal explanations in other areas of natural science.

When the contributions of the fourteenth-century Mertonians to the science of mechanics are assessed, these similarly show an extensive development of the mathematical tools that were to make seventeenth-century mechanics a possibility, but an almost complete neglect of the experimentation and measurement that would be necessary for its realization. Thomas Bradwardine, as we have seen, originally addressed himself to a problem in dynamics when he attempted to relate the speed of motion to the forces and resistances involved in its production. Those who were inspired by his work, however, never seriously addressed themselves to the problems of the efficient causality involved in local motion, but turned instead to an extensive development of kinematics in terms of the spatio-temporal relationships associated with motions of all types. This involved them in works of great mathematical sophistication, but which could be treated almost entirely in terms of imaginary examples and never seemed to require verification through a study of natural motions. Although lip service was paid to causes as explanatory factors, the scientific ideal to which the Mertonians subscribed did not involve a search for physical causes in the methodological sense of the *Posterior Analytics*. As a consequence, and this we have argued elsewhere, the mentality of the experimenter, who attempts to duplicate natural effects by simulating or initiating their causes in a controlled situation, is not to be found in the Mertonian contribution.[123]

Therefore, if we are to regard modern science, on the model of its seventeenth-century understanding, as involving both mathematical reasoning and experimentation, we shall have to conclude that the beginnings of modern science at Oxford University in the thirteenth and fourteenth centuries fostered mainly the mathematical component. The Oxford effort led also to a certain amount of idealization and to an over-simplification of optical and mechanical problems — a step that was undoubtedly necessary at this stage of science's development. Without such simplification, in fact, as the extraordinary complexity of later analyses of such commonplace phenomena as the fall of bodies in the earth's atmosphere has shown, no beginning would have been made at all. Yet the systematic use of experimentation to make this type of analysis possible would require a still more extensive theoretical development, and it would receive inputs from sources other than Oxford University. To investigate these sources we must now turn our attention to universities on the Continent where Aristotelian thought was also stimulating a scientific renaissance, and first and foremost to the University of Paris.

CHAPTER THREE

Medieval Science at Paris

THE Aristotelianism that arrived in the universities of the Latin West in the thirteenth century included much that was not Aristotle. One reason for this may be traced to the peculiar order in which the various books of the *corpus aristotelicum* became available, prompting some of their early readers to give premature interpretations in the light of Augustinian principles. Another reason is the only gradual disengagement of Aristotle's thought from the Platonic and Arabic overtones it had acquired through translation from the original and intermediate languages. Most important of all, however, was the mistaken inclusion of the *Liber de causis* in this *corpus,* which led practically all thinkers up to Thomas Aquinas — who recognized the work as that of Proclus — to regard it as the most basic part of Aristotle's metaphysics.[1] Since it was at Paris that the true identification of this work was finally made, it is fitting that there also the first attempts were made by Aquinas and his mentor, Albert the Great, to teach a purer Aristotle and to see how the heterodox content of his teaching could be reconciled with Christianity.

From the viewpoint of the history of science, the return to this purer Aristotle necessitated a withdrawal from the Neoplatonic mathematicism so noticeable in the work of Grosseteste and the development of a more empirical approach to the world of nature. It is this changed attitude, as we shall argue in the present chapter, that resulted in a stronger commitment to realism at Paris than at Oxford, which was able to withstand the influx of nominalist views in the fourteenth century and at the same time adapt the kinematical insights of the Mertonians to new solutions of dynamical problems. In what follows attention will be paid mainly to the ways in which changing views of causality were influential in this process, and the exposition will be divided into three phases. The first will discuss at some length the distinctive interpretation given to Aristotle through the efforts of Albert the Great and Thomas Aquinas; the second will analyze two experimental contributions that were made in the context of causal analysis, namely, those of Peter of Maricourt and Theodoric of Freiberg; and the third will detail the fourteenth-century advances in mechanics made by the group later known as the Paris Terminists.

1. *Albert the Great*

The position of St. Albert the Great with respect to the University of Paris parallels in some ways that of Robert Grosseteste with respect to Oxford University.[2] Albert had a distinct advantage in that far more sources were available to him, including the translations and commentaries of Grosseteste himself, and he worked indefatigably at understanding these newly available materials. His status as a master in the Dominican Order also enhanced his productivity, since he was requested by his fellow friars to make all of Aristotle's science available to them in Latin. Rather than comment on the individual works, Albert chose to paraphrase the entire body of Aristotelian teaching, including the tradition of the commentators both Greek and Arabic, but with a particularly careful analysis of Avicenna and Averroës. So monumental was this work that it earned for him great admiration in the schools, where he was called "Doctor universalis" and was even referred to as "the Great" in his own lifetime.[3]

Albert's exposition of the *Posterior Analytics* is not as interesting as Grosseteste's, mainly because it is only a summary of the broad lines of Aristotle's thought and has no detailed consideration of examples. There can be little doubt, however, that Albert consulted Grosseteste's commentary when composing his paraphrase. Augustine Borgnet, who edited the Latin text of Albert's work in 1890, identifies six places where the doctrines of "Lincolniensis" are being discussed,[4] and undoubtedly more would be obvious to anyone with a detailed knowledge of Grosseteste's commentary. There may be some connection, for example, in the fact that Albert includes among his division of propositions those referred to as *propositiones experimentales,* and gives as an example of this type of proposition that scammony purges red bile.[5] Again, Grosseteste's long discussion of the ways in which universals are incorruptible is noted by Albert and then contrasted with the theories of Averroës, Alfarabi, Themistius, and Alexander of Aphrodisias.[6] In this context Albert's own answer to the way in which eclipses are incorruptible is worth remarking. He says that an eclipse can be considered as a property and in relation only to its subject, the moon, and then it does not always exist. Alternatively it can be considered in relation to its subject taken jointly with the cause that produces it, i.e., the sun and moon in their proper spatial configurations, and then a universal eclipse always exists, since it is necessarily produced whenever this arrangement occurs. The property considered only in itself, however, is neither necessary nor eternal.[7]

Albert's treatment of the distinction between *quia* and *propter quid* demonstration is unexceptional, although he does discuss the cases where physics is subalternated to mathematics and notes that if the properties of natural things are purely quantitative the mathematician can give the *propter quid* explanation, whereas if they are associated with the nature of the substance being dealt with, then "the *propter quid* is not given by the superior science but by the inferior," i.e., by physics.[8] Albert's analysis of the thunder and lightning example is unfortunately limited to three different explanations of the ways in which definitions can be related to demonstrations; he makes no attempt to identify any of the causes of these phenomena treated so imaginatively by Grosseteste.[9]

Throughout his paraphrase of the *Posterior Analytics* Albert refrains from noting differences of opinion between himself and the Oxford master. As he moves on through his expositions of the *Physics* and *Metaphysics*, however, and particularly as he considers in more detail the way in which mathematical principles are related to natural things, it becomes clear that Albert's views on this subject differ from those of the Oxford school. The point of contention is their Neoplatonism, and the rejection of this inevitably has repercussions on Albert's understanding of the role of experience and empirical findings in the development of natural science. His paraphrase of the *Physics* is the earlier work and is quite encyclopedic in nature, showing Albert's skill more as a synthesizer than as an original expositor. In explaining Book 3, for example, on the question as to how motion is to be classified among the Aristotelian categories, he essentially summarizes the views of Avicenna and Averroës on the nature of motion — though while so doing he makes a number of distinctions that adumbrate the controversy over *forma fluens* and *fluxus formae* that was to arise in the fourteenth century.[10]

It is in his exposition of the *Metaphysics,* however, that Albert sets aside his role of encyclopedist and launches out against the *error Platonis*, "who said that the principles of natural being are mathematical, which is completely false."[11] It is not Plato, of course, who is his target, but rather those whom he refers to as the *amici Platonis,* and whom Weisheipl has identified as Grosseteste, Robert Kilwardby, and Roger Bacon.[12] Kilwardby, as Albert's fellow Dominican, was probably the principal adversary he had in mind. Himself a secular master at Paris for some time before entering the Dominican Order, during which period he taught Roger Bacon, Kilwardby was sent by the Order *c.* 1245 to study theology at Oxford.[13] Here he came under Grosseteste's influence, and shortly thereafter composed a lengthy treatise, *De ortu scientiarum,* which explains the nature, origin, and division of most of the sciences then known.[14] Fundamental to his treatment of the speculative sciences is his teaching on intellectual abstraction, which, while based on Aristotle, incorporates a number of Platonic insights. All of reality, in Kilwardby's view, is structured in terms of three classes of forms: perceptible qualities, dimensive quantities, and intelligible substances. The abstraction that is characteristic of physical science is one that leaves aside all individual characteristics to study the nature of perceptible, extended substances. Mathematical abstraction goes beyond this, leaving aside perceptible qualities and studying only extended substances, whereas metaphysics abstracts even from extension and has for its object pure substance alone. Kilwardby's theory of abstraction was thus based on what he thought was the priority of forms actually existing in nature: first and most fundamental was the unitary form of substance, then came number and extension, and finally sensible qualities and individuality. Following this insight further, Kilwardby maintained that any of the later forms could be explained in terms of the antecedent forms, so that since all perceptible qualities and sensible phenomena are rooted in the dimensionality of bodies, the proper explanations for them would be found in mathematics. Thus natural science is intrinsically, as it were, subalternated to mathematics, and mathematics in turn is similarly subalternated to metaphysics, since the basic principle of mathematics, the unit, is in reality the unity of intelligible substance.[15]

Albert never mentions Kilwardby or the "Oxford Platonists," as Weisheipl has termed them, by name, preferring as we have seen to attribute their doctrine to Plato, whose "error we have rejected in the books of the *Physics,* and which we shall again reject in the following books of this science [i.e., metaphysics]." [16] Thus it is that in the *Metaphysics* Albert discourses extensively on this subject to show (1) that the principles of natural science are not mathematical; (2) that the object of mathematics is not an ontologically antecedent form, but rather an abstracted entity; and (3) that the unity which is the principle of number is essentially different from the unity that is identical with being.[17] Albert grants that metaphysics is the last science to be studied, and that it presupposes all of the other sciences, but in no way does he feel that it makes them superfluous. A person cannot know the individual things of nature merely by contemplating mathematical or metaphysical principles; he must study them in their proper natures with their own distinctive principles, and for this the knowledge of all the natural sciences is required. For Albert, then, neither the mathematician nor the metaphysician has a privileged insight. The world of nature will yield its secrets only to the person who studies it carefully through observation and empirical reasoning.

That this was more than a speculative conviction for Albert can be seen from his entire life's work, which resulted in empirical contributions to practically all of the sciences and made him the most celebrated precursor of modern science in the thirteenth century. The details of this work have been treated elsewhere, and are readily available in standard histories of science.[18] What perhaps needs underscoring is that Albert conducted his researches with a conscious awareness that he was searching for the causes of natural things. "In natural science," he writes, "we do not investigate how God the Creator operates according to His free will and uses miracles to show His power," — this following a remark on the *Timaeus* — "but rather what may happen in natural things on the ground of the causes inherent in nature." [19] In his *De mineralibus* he insists that "it is not for natural science to simply accept what has been said, but to look for the causes in natural things." [20] His understanding of causal inquiry, therefore, was not limited to searching other authors for their explanations of natural phenomena, but stressed rather an independent study of the available facts. This attitude led him to question even the authority of Aristotle, as in the *Libri Meteororum*, where he observes that "Aristotle must have spoken from the opinions of his predecessors and not from the truth of demonstration or experiment." [21] And again: "Whoever believes that Aristotle was a god, must also believe that he never erred. But if one believes that he was human, then doubtless he was liable to error just as we are." [22] Albert was firmly convinced, also, that the principles of natural science must be rooted in sense experience. So he states in his exposition of the *Physics:*

> Anything that is taken on the evidence of the senses is superior to that which is opposed to sense observation; a conclusion that is inconsistent with the evidence of the senses is not to be believed; and a principle that does not accord with the experimental knowledge of the senses (*experimentali cognitioni in sensu*) is not a principle but rather its opposite.[23]

Like Grosseteste and Bacon, Albert speaks often of "experimental knowledge" and "experiment," meaning by these expressions careful generalizations made on the basis of repeated experiences. "Much time is required for an experiment to be verified, that it not be defective in any respect," writes Albert, quoting Hippocrates in corroboration of this. "In fact," he continues, "an experiment should not be verified in one way only, but under all different circumstances, so that it may provide a certain and correct principle for [future] work."[24] Then he explains that this is why youths readily grasp the fundamentals of mathematics, but find physics more difficult from the very fact that they lack the requisite experiential knowledge.[25]

As has been noted in the cases of Grosseteste and Bacon, one should guard against reading too much into statements of this type, particularly by way of seeing in them an anachronistic use of experimental method as developed in seventeenth- and eighteenth-century science. Certainly there is enough of the fabulous in Albert's immense literary output to serve as a warning in this regard.[26] But at the same time one should not fail to notice that Albert was committed, in principle and in fact, and in a way conspicuously absent in most of his contemporaries, to an empiricist program. His studies of the thermal effects of sunlight, his observations of fossils unearthed in excavations, his comparative studies of plants and animals (which show a remarkable sense of morphology and ecology), his tracing of the development of the chick embryo by opening eggs at successively longer time intervals, his descriptions of insect mating, and his contributions to taxonomy in general, all serve to confirm that his was no facile, ivory-tower approach to nature, but rather one very much in the spirit of experimental science.[27]

2. Thomas Aquinas

Albert the Great's observational and empiricist techniques bore fruit among his disciples and those who came under their influence.[28] Of yet greater importance, however, was the general methodological impact on his most celebrated student, St. Thomas Aquinas, whose monumental contributions to theology overshadow his other intellectual efforts, particularly those in the natural sciences. Although concerned mainly with the supernatural, Aquinas was convinced that one could only approach this realm through a knowledge of the world of nature. This conviction apparently grew stronger the more he worked on his masterpiece, the *Summa theologiae*, so much so that toward the end of his life he turned away from this work while it was yet unfinished to comment on the *libri naturales* of Aristotle. He probably completed his exposition of *De caelo et mundo*, one of his best works as a commentator, during the academic year 1272–73 at Naples, and he ceased commenting on *De generatione et corruptione* and the *Meteorologica* only late in 1273, and this partly for reason of failing health, three months before his death. Earlier he had commented carefully on Aristotle's *Physics, De anima, De sensu et sensato*, and *De memoria et reminiscentia*, and had written the brief letters *De motu cordis* and *De occultis operibus naturae*. What typifies all of these works is his concern with methodology, the proper order of questioning, and particularly the logical structure underlying Aristotle's more or less cryptic exposition.[29]

Such a preoccupation could only spring from a thorough knowledge of the *Posterior Analytics*, a knowledge that is much in evidence in Thomas's exposition of the same, written *c.* 1270. Unlike Albert's paraphrase, this is a detailed commentary, even more so than the Great Commentary of Averroës, which Aquinas took into account in his own exposition. Faithful to his teacher, Aquinas was alert to the problems posed by Neoplatonism in understanding the relationship of mathematics to physics, and was careful to analyze all of Aristotle's examples, particularly those as we have seen that perplexed Grosseteste, in a way that preserves the autonomy of the physical sciences.[30] His great contribution in this endeavor was his explanation of the way in which all four causes can be used when demonstrating in natural science, and particularly how final causality is to be employed in the demonstrative process. It is this insight that enabled Aquinas to maintain that there can be true science of corruptible and changeable things, and thus answer the basic problem posed by Plato, which continued to perplex commentators on Aristotle until Thomas's day.

Aquinas's position on this important issue becomes clear from his comment on Aristotle's statement that "no attribute can be demonstrated or known by strictly scientific knowledge to inhere in perishable things," and on his additional observation that "the same is true of definitions." Here he digresses somewhat, as follows:

> For a better understanding of this passage it should be noted that it is possible to give different definitions of the same thing, depending on the different causes mentioned. But causes are arranged in a definite order to one another: for the reason of one is derived from another. Thus the reason of matter is derived from the form, for the matter must be such as the form requires. Again, the agent is the reason for the form: for since an agent produces something like unto itself, the mode of the form which results from the action must be according to the mode of the agent. Finally, it is from the end that the reason of the agent is derived, for every agent acts because of an end. Consequently, a definition which is formulated from the end is the reason and cause proving the other definitions which are formulated from the other causes.[31]

Having thus set up the "end," or final cause, as the principal defining and explanatory factor in natural things — for these alone, as opposed to mathematical and metaphysical entities, are generable and corruptible — Aquinas turns his attention briefly to Plato to note that it was Plato's concern over the fact "that demonstration, as well as definition, is not of destructible but of eternal things," that led him to posit his theory of Ideas.[32] He notes also that Aristotle, thinking Plato's theory too extreme, held that there *could be* demonstration of destructible things, for "although those sensible things are destructible as individuals, nevertheless in the universal they have a certain everlasting status."[33] Thus Aquinas reads Aristotle to mean that "demonstration bears on [such] sensible things universally and not individually."[34] To explain how this can be, Aquinas has recourse to what had by this time become the paradigm of all scientific demonstration, the example of the eclipse of the moon, and proceeds to give his exegesis of this cryptic illustration in the Aristotelian text:

The moon is not always being eclipsed, but only now and then. Now things that occur frequently, so far as they are such, i.e., so far as demonstrations are given concerning them, are always; but they are particular, so far as they are not always. But demonstration cannot be of particulars, as we have shown, but only of universals. Hence it is clear that these things, insofar as there is demonstration of them, are always. And as in the case of the eclipse of the moon, so in all kindred matters.[35]

The last sentence, as we shall see, is Aquinas's opening wedge for applying the methodology of this astronomical example to other examples in what was for him the purely physical science of the sublunary region. Under "all kindred matters" he includes natural phenomena that are of frequent occurrence but are not eternal and immutable, as are the subject matters of mathematics and metaphysics. This becomes clear in what follows:

However, there are certain differences to be noted among [these cases]. For some are not always with respect to time, but they are always in respect to their cause, because it never fails that under given conditions the effect follows, as in the eclipse of the moon. For the moon never fails to be eclipsed when the earth is diametrically interposed between sun and moon. But others happen not to be always even in respect to their causes, i.e., in those cases where the causes can be impeded. For it is not always that from a human seed a man with two hands is generated, but now and then a failure occurs, owing to a defect in the efficient cause or material cause. However, *in both cases the demonstration must be so set up that a universal conclusion may be inferred from universal propositions by ruling out whatever can be an exception either on the part of time alone, or also of some cause.*[36]

Here we see how Aquinas has built upon the interpretations of the eclipse example given by Grosseteste and by Albert the Great.[37] He not only has gone further to include natural, as opposed to astronomical, examples in his analysis but he has also signaled, in the italicized sentence, that a uniform methodological procedure must be used when one seeks to demonstrate universal conclusions in cases of frequent occurrence.

This procedure is explained by Aquinas more fully in other contexts, where he refers to it as demonstration *ex suppositione,* and is careful to point out the precise ordering of causes that is essential to demonstration of this type.[38] Briefly, the procedure is the following. "On the supposition" that any effect is to be produced, as the end or final cause of a natural process, one may be able to demonstrate *propter quid* the efficient and material causes required to produce that effect, and if so, such reasoning will produce truly universal knowledge even though the effect is not always occurring. Thus one who knows astronomy can state in precise detail all that is required to produce an eclipse of the moon, and this will be universally true whenever and wherever an eclipse is to occur. The same method is applicable in the sublunary world, although here it is more difficult to exemplify; possibly for this reason Aquinas's instance of the "man with two hands" is proposed only at a level of great generality. He returns to this type of reasoning, however, when discussing the statement with which Aristotle begins Chapter 30 of Book I, namely, "there is no knowledge by demonstration of chance conjunctions." [39] Here Aquinas wishes it to be well understood that there is a difference between events that occur by mere chance or fortune and those that occur with some degree of regularity, though not always. He writes:

> It should be noted, however, that there happens to be demonstration of things which occur, as it were, for the most part, insofar as there is in them something of necessity. But the necessary, as it is stated in *Physics II,* is not the same in natural things (which are true for the most part and fail to be true in a few cases) as in the disciplines, i.e., in mathematical things, which are always true. For in the disciplines there is *a priori* necessity, whereas in natural science there is *a posteriori* (which nevertheless is prior according to nature), namely, from the end and form. Hence Aristotle teaches there that to show a *propter quid,* such as, if this has to be, say that if an olive is to be generated, it is necessary that this, namely, the olive seed, pre-exist, but not that an olive is generated of necessity from a given olive seed, because generation can be hindered by some defect.[40]

Of course, the generation of an olive, or of a man with two hands, was not understood in any great detail by Aristotle or by the medievals; they thought perhaps that they could do better when explaining eclipses, rainbows, and even thunder and lightning. In all of these cases, however, it was the end or form regularly attained in nature's operation that Aquinas recognized as the essential starting point for this type of demonstration. It is well to insist on this, because the modern reader might be tempted to understand the expression *ex suppositione* as a merely conjectural explanation proposed in a dialectical way as within the realm of logical possibility. Aquinas's procedure should certainly not be taken as hypothetical in this sense; rather it had to be rooted in an observational or an empirical approach to nature that was quite consistent with the teaching of his mentor, Albert the Great. For it is only a patient study of natural processes that enables one to detect the regularities and uniformities, later to be designated as laws of nature,[41] that form the empirical basis for scientific explanations.

Aquinas remains concerned, throughout his commentary on the *Posterior Analytics*, with epistemological problems of this type, particularly those relating to universals and the possibility of attaining scientific knowledge through sense observation. He thus addresses himself to Grosseteste's problem that the universal "is not restricted to the here and now" but must be "always and everywhere," and solves it in terms of his distinctive theory of knowledge, wherein the universal is obtained from singulars by a process of intellectual abstraction.[42] He writes:

> . . . if to exist always and everywhere pertains to the [universal's] very notion . . . it would follow that nothing would be universal if it were not found everywhere or always. According to this, "olive" would not be universal because it is not found in all lands. Hence, the statement under consideration must be understood after the manner of a negation or abstraction, i.e., that the universal abstracts from every definite time and place. Hence of itself, just as it is found in each thing in one place or time, so it is apt to be found in all.[43]

Moreover, just as the universal, in this view, is understood as abstracting from any particular instance, so in Aquinas's theory of knowledge it is also obtained by the abstractive power of the human mind. Following Aristotle, however, Aquinas stresses that there is no essential connection between our observing a particular event and our coming to a scientific knowledge of that event.[44] "Rather," he writes, "we receive universal knowledge from many individually observed cases in which the same thing is found to happen."[45] Yet sometimes, he admits, a single careful observation may be sufficient to generate a scientific explanation.[46] In these and in other texts Aquinas generally steers a judicious course between the extremes of naïve phenomenalism and a precipitate rationalism, while arguing for the possibility of valid generalizations in the physical sciences.

From his account of the relationships between *quia* and *propter quid* demonstrations in the same science, and from other *loci* where he is discussing the type of demonstration to be found in physical science, there can be no doubt that Aquinas, like Albert, thought that strict *propter quid* demonstration was possible in such science, provided the proper methodological procedure was used in its attainment. Aquinas discusses the examples we have already seen, i.e., the planets' twinkling and the moon's phases, the rainbow's properties and the slow healing of a circular wound. Apparently he subscribes to Grosseteste's explanation that "the fixed stars twinkle because in gazing at them the sight is beclouded on account of the distance."[47] He warns, however, that one should be sure that he has a convertible effect before trying to convert any *quia* demonstration into a *propter quid*. Thus it is all right "if one proves that Venus is near because it does not twinkle," whereas "one cannot conclude universally of stars that they are near because they do not twinkle."[48]

Aquinas's understanding of the subalternation of the sciences, as is clear from his discussion of the rainbow, is along Albert's lines and not Kilwardby's. Rather than have the mathematician be concerned with an underlying geometrical structure in matter, such as might be the case for anyone subscribing to Grosseteste's "metaphysics of light," Aquinas insists that the mathematician is concerned only with abstractions of a special type, but that he is nonetheless able to apply these fruitfully to the solution of physical problems. He explains:

> Although the items which geometry considers exist in matter, for example, the line, plane and so on, nevertheless geometry does not consider them precisely as they are in matter, but as abstracted. For those things that are in matter according to existence, geometry abstracts from matter according to consideration. Conversely, the sciences subalternated to it accept those things which were considered in the abstract by geometry and apply them to matter. Hence it is plain that it is according to the formal cause that geometry states the *propter quid* in those sciences.[49]

The formal cause involved is thus one of quantitative form, but this as abstracted and not as in some way latent within matter; it is in terms merely of the *application* of such forms that one is to understand the scientific character of geometrical optics and its success in explaining the properties of the rainbow.[50]

While recognizing this type of *propter quid* explanation through a subalternating science, Aquinas understands that *propter quid* demonstrations are best given in any one science in terms of the four causes proper to that science, whether one cause be used to demonstrate another according to the order already indicated, or whether a proper attribute be demonstrated through the definition, i.e., through the *quod quid* as made up of any or all of the causes that constitute the essential nature of the thing.[51] Thus, following along in Aristotle's text, he discourses in detail on how it is possible to obtain *propter quid* demonstrations in the four genera of causes, using these to manifest both the *quod quid* in its various essential elements and also the proper attributes.[52] One final example, related to such a use of demonstration through the final cause and concerned with the definition of thunder, is of interest because of the difference it highlights between Aquinas's approach and Grosseteste's. Aquinas points out that there are two ways of arguing based on the final cause: (1)

one is to go from a cause that is "pre-existing," i.e., from a material or efficient cause, and argue from this to the final cause as the effect that is to be produced; (2) the other is to argue from "a posterior cause which is posterior in the order of becoming," i.e., the final cause as this would be taken *ex suppositione*, as already explained.[53] It is only in this second way, as we have seen, that necessary demonstrations can be obtained in the natural sciences, because of possible defects in the matter or in the operation of efficient agents. But even in this second way, there are two further alternatives in considering the final cause: (a) one as the form that is intrinsic to the generating process, and which ends or terminates that process; and (b) the other as some further end which results from the completed form, once it has been produced. These two subdivisions can be seen in the example of thunder: (a) if thunder is to be produced, and "if it is quenched fire, it is necessary that it hiss, i.e., make the sound and roar of fire being quenched," and when it does this, thunder is produced; or, alternatively, (b) if, according to the Pythagoreans, "thunder takes place to strike terror into the denizens of Tartarus, then one should say that thunder takes place to the end that the men in Tartarus shudder."[54] The implied distinction between the intrinsic and the extrinsic final cause is quite important for understanding Aquinas's methodological advance over his contemporaries.[55] Grosseteste, as we have noted, assigns the final cause of thunder as nature's attempt "to purify the air and separate out extraneous qualities."[56] This, like terrifying those detained in the infernal regions, is extrinsic to the process that produces thunder. Such extrinsic final causes are extremely difficult to discover in nature and generally elude human observation; this, seemingly, is why Aristotle states that "the final cause is least obvious where matter predominates,"[57] and Albert the Great, in his commentary on the *Posterior Analytics*, remarks that "for natural attributes there is no demonstration through the final cause."[58] With regard to the intrinsic final cause, however, this is regularly manifested in nature as the form at which natural processes terminate, whether such processes be productive of entities like olives or of phenomena like the rainbow and thunder. An empirical approach can reveal the intrinsic final cause, which can then be the basis for demonstration of all the other causes, through the use of the *ex suppositione* procedure already outlined. It is this technique, for Aquinas, that can yield *propter quid* demonstrations that are proper to natural science and that result from its proper methods. Thus he did not have to have recourse to the "metaphysics of light" or to any other privileged insight to justify the possibility of a natural science. And he could even, in so doing, lay claim to the Aristotelian character of such science as a perfect form of knowing (*scire simpliciter*), despite its being concerned with things that do not happen with absolute necessity, but only for the most part.

a. Mathematics and Proof

As we have seen, neither Albert nor Aquinas saw mathematical form as existing in physical things antecedently to sensible quality and to individuality; unlike Kilwardby they understood it as form abstracted from an existing physical entity by a mental process that leaves aside all irregularities and individual characteristics resulting from matter and motion. They also recognized that quantity could be a legitimate subject of investigation in two sciences, namely, natural science and mathematics, although it would be considered differently in the two.[59] For the natural scientist mathematical form is not the most basic; more primary, for him, is the physical nature of the entity being studied, for it is this nature that dynamically takes on the form or shape by which one recognizes the species and readily distinguishes, for example, a man from a lower animal. The quantitative characteristics that so serve to identify a natural species are not themselves mathematical, but physical. Yet it is such physical entities, existing in different number and quantitative dimensions, that are the originative sources of the idealized forms studied by the mathematician. For Albert and Aquinas, therefore, the insight afforded by mathematics is not deeper or more "divine," as the Platonists would have it, but is more superficial than a physical insight. As a consequence, a study of the mathematical features of a physical entity does not necessarily explain its nature, although it can accurately describe the quantitative characteristics of that nature, and it may help in discovering a physical explanation for them. Since, moreover, what is true of mathematical quantity is verified also of physical quantity, a subalternated science such as mathematical physics can give *propter quid* demonstrations of conclusions that pertain to physics through the use of a mathematical middle term.[60]

This view of the relationship of mathematics to physics is central for understanding the advance made at Paris over the methodology of the thirteenth-century Oxonians. The theory of knowledge behind it is far from simplistic, and it actually allows for a multiplicity of uses of mathematics in the physical sciences, at least two of which should concern us here. The first, as we have seen, is the employment of a mathematical middle term to give a *propter quid* explanation, which in turn may stimulate the search for the underlying nature and yield conclusive proof in terms of proper physical causes. The second — and more interesting, as it will turn out, from the viewpoint of modern science — is a dialectical rather than a demonstrative employment; it functions at the level of a formalistic explanation or hypothesis that would "save the appearances," but would not necessarily correspond to the physical reality underlying them, although it too could assist in attaining knowledge of that reality. The difference between the two can be made clear by the analysis of examples, which may conveniently be taken from Aquinas's late commentary on the *De caelo et mundo*.

An example of the first procedure is suggested by Aquinas's methodological statement that "the astronomer and the physicist both prove the same conclusion — that the earth, for instance, is round: the astronomer by means of mathematics (i.e., abstracting from matter), but the physicist by means of matter itself." [61] Here he obviously has in mind the arguments offered by Aristotle in Chapter 14 of Book II of the *De caelo* to prove the earth's sphericity, and thus we should look in this *locus* for Aquinas's detailed understanding of these arguments. We find here that he stays close to Aristotle's text but at the same time that he adds his own insights and incorporates data that were not known in Greek antiquity. He thus discerns three astronomical arguments that employ mathematics to show that the earth is round, the first two geometri-

cal and the third mensurational: the former are taken from
"what appears according to sense" in the eclipse of the moon
and in the appearance of the stars when seen from different
points on the earth's surface; the latter "is based on measurements of the earth." [62]

The first argument is of interest for what it reveals of
Aquinas's knowledge of projective geometry. Thus he states:

> Unless the earth were spherical, an eclipse of the moon
> would not always reveal circular segments; for we ob-
> serve that whenever the moon is eclipsed, its dark and its
> shining portions are distinguished by a curved line. Now
> an eclipse of the moon results from its entering the
> earth's shadow — hence the earth's shadow appears to be
> round. From this it appears that the earth, which makes
> such a shadow, is round — for only a spherical body is
> apt always to cast a round shadow.[63]

Aquinas goes on to consider the effect of the sun's size (larger
than, equal to, or smaller than the earth) on the projection
and concludes: "Now, no matter which of these it should be,
it would follow, on account of the earth's sphericity, that its
shadow would cut the moon according to a circular line." [64]
He also contrasts the appearance of the lunar eclipse with the
phenomenon of the moon's phases, to which allusion was
made in the discussion of Grosseteste, as follows:

> Now someone could say that this circular section is not
> due to the earth's rotundity but to the moon's. But to ex-
> clude this [Aristotle] adds that in the monthly waxing
> and waning of the moon, the section of the moon takes
> all differences of shape — for sometimes it is divided by a
> straight line, as when it is divided through the center, for
> example, on the 7th and the 21st days; at other times, it
> is *amphicurtos* [gibbous], having a circular or arc-like
> section, namely from the 7th to the 21st; at still other
> times it is concave [crescent], as from the 1st to the 7th,
> and from the 21st to its total waning. All this happens
> according to its positions in relation to the sun, as has
> been said above. But during eclipses the line dividing
> the moon is always "gibbous," i.e., circular. Since, there-
> fore, the moon is eclipsed by the interposition of the
> earth, the rotundity of the earth, since it is spherical, is

the cause of such a shape with respect to the division of the moon.[65]

The second argument is based on the stellar appearances, and particularly the different elevations of the celestial pole when "one moves to the north or to the south," which also makes a difference in the stellar horizon. Aquinas reports the observational data as follows:

> Since, therefore, because of the difference of horizon, in northern lands the north pole is higher and the opposite pole is lower, it happens that certain stars which are near to the antarctic pole are not perpetually hidden, but are sometimes seen in lands more to the south, for example, in Egypt and about Cyprus, which are never seen in the more northerly region. Conversely, certain stars are always visible in the more northern regions, which in more southern regions are hidden by setting.[66]

Using these data, Aquinas shows how they may be used geometrically to establish that the earth is a sphere:

> And from this it appears that the earth is rotund in shape, especially according to its aspect at the two poles — for if it were flat, all those dwelling on the whole face of the earth to the south and north would have the same horizon, and the very same stars would appear to them and be hidden from them, no impediment arising from the bulge of the earth.[67]

At this point Aquinas adds another argument that anticipates those offered by Copernicus in his *On The Revolutions of the Celestial Spheres:*

> And with a similar argument it is proved that the earth is round toward the east and west — otherwise no star would arise any earlier for people in the east than for those in the west. For if the earth were concave, a rising star would appear first to people in the west; but if the earth were flat, it would appear to everyone at the same time. But it is evident that a rising star appears first to those in the east, if we consider a lunar eclipse. If such an eclipse appears in a more easterly region about midnight, it will appear before midnight in a more westerly region, depending on the amount of the distance. From

this it is plain that the sun rises earlier and sets earlier in a more easterly region.[68]

Aquinas's final mathematical argument for the shape of the earth, following Aristotle, is based on the attempts of various "mathematicians" to calculate the earth's circumference. Aquinas notes Aristotle's figure and that of Simplicius, and refers to "the more careful measurements of present-day astronomers," for which he cites the Arab Alfragani. "From all this," he concludes, "we can argue that the earth's quantity is not only spherical, but not large in comparison to the sizes of the other stars." [69]

The foregoing arguments, it should be noted, are all based on metrical aspects of the appearance of the heavens as seen from the earth, which permit of arguments a posteriori, i.e., from effect to cause, to show that the earth must be a sphere. Such arguments can be the cause of one's recognizing the earth's sphericity, but they do not explain how the earth came to be a sphere; for this it is necessary to proceed a priori, from cause to effect, considering the matter of which the earth is made and the motion, or tendency to motion, that is characteristic of its elemental constituents. It is the physicist, rather than the astronomer or mathematician, in Aquinas's view, who must discover such an explanation. This need not exclude quantitative considerations, since the physicist's proper subject matter embraces physical quantity, as we have seen; yet he ought not to stop with these but should proceed beyond them to a knowledge of the physical cause that produces the characteristic shape. Again following Aristotle, Aquinas sees in the text two arguments that make this cause apparent, one from the kind of motion that characterizes the earth and its parts and the other from the "figure" or spatial orientation of that motion.[70]

Aquinas's summary of the first line of reasoning is that the earth has a spherical shape simply because each of its parts "is naturally moved to the center of its gravity." Aristotle, he notes, proposed the proof in the context of his predecessor's belief "that the earth was newly generated by its parts coming together from every direction toward the center," and that this motion was "caused by the violence of the heavens' gyration." In Aquinas's own opinion, "it is better and truer to suppose that this motion of the parts of earth occurs naturally, on account of these parts having a gravitational tendency toward the center." [71]

The second argument makes use of this same line of reasoning, but considers a quantitative modality of the motion of parts tending toward the center that gives a confirmatory proof of the earth's sphericity. As Aquinas explains it:

> [Aristotle] says that all heavy bodies, from whatever region of the heavens they are moved, are carried to the earth "at like angles," i.e., at right angles formed by the straight line of the body's motion with a line tangent to the earth — which is evident from the fact that heavy objects do not stand firmly on the earth unless they are perpendicular to it — but that heavy bodies are not carried to the earth "side by side," i.e., in parallel lines. Now all this is ordered to making the earth spherical by nature: for heavy bodies have a like inclination to the place of earth no matter from what part of the heavens they are released, and thus there is an aptitude for additions to the earth to be made in a like and equal matter on all sides, which makes it to be spherical in shape. But if the earth were naturally flat on its surface, as some used to claim, the motions of heavy bodies from the heavens to earth would not be from all sides at similar angles. [72]

It is this universal mode of gravitation that makes the earth be "spherical by nature," and thus the proof is through natural principles, i.e., from matter and from motion, even though it concludes to a quantitative modality, i.e., sphericity, that is the effect of uniform gravitational tendency.

It should be noted in this proof that the physical cause assigned need not make the earth a perfect sphere. "The bulges of mountains and the depressions of valleys" arise, Aquinas concedes, although "not of noticeable dimensions in relation to the whole earth," and he attributes them to "some other incidental cause." This confirms what has been said above, in the sense that, for Aquinas, pure or perfect mathematical shape does not exist in physical reality. It is only the human mind, abstracting from such irregularities as mountains and valleys, that can conceive the earth as a perfect sphere. And yet Aquinas insists that since it has a natural tendency to be spherical, regardless of the incidental causes that result in departures from this shape, there is no reason why it should not be referred to as a sphere.[73]

These arguments, it may be noted, although offered with respect to the shape of the earth, can be extended to the moon and other heavenly bodies, as we have seen intimated in Grosseteste's commentary on the *Posterior Analytics*. Such instances show how the medievals, following Aristotle, had no hesitancy in applying mathematics to physical problems, whether this was merely to yield knowledge of the fact, as in the first line of reasoning, or knowledge of the reasoned fact as in the second, which assigns a physical cause but at the same time concludes to a mathematical modality that is the result of the cause's operation.

b. Mathematics and Hypothesis

Apart from these demonstrative uses of mathematics in physical science, Aquinas was also aware of a more hypothetical or dialectical use that would merely "save the appearances." [74] He resorts to this when faced with the serious difficulty of reconciling the strictly homocentric theory of the universe of Eudoxus and Calippus with the modified geocentric theory of Ptolemy. In his earlier writing Aquinas favors the Eudoxian system, but he also mentions the Ptolemaic system at least eleven times and with increasing frequency in his later works.[75] It is not surprising, then, that he finally expresses himself on the relative merits of these alternative theories of the structure of the universe. He does so in his commentary on the *De caelo*, where he begins with "a fact which is sensibly evident," namely, that "we see the stars and the whole heavens moving," and notes that "this must be caused either by the motion of the thing seen or of the one seeing" [76]—an implicit acknowledgment of the basic relativity involved in accounting for the appearances. He traces the history of the mathematical constructions of Eudoxus and Calippus and their improvements by Hipparchus and Ptolemy, noting how the last named corrected the sequence of the planets, knew of the precession of the equinoxes, and successfully explained the motions of the planets by eccentrics and epicycles.[77] It is in the context of his discussion of these various attempts made by astronomers to account for anomalies or irregularities such as these that "appear with respect to the motions of the planets," that Aquinas makes the following statement:

> Yet it is not necessary that the various suppositions which they hit upon be true — for although these suppositions save the appearances, we are nevertheless not obliged to say that they are true, because perhaps there is some other way men have not yet grasped by which the stellar appearances can be saved.[78]

An evaluation similar to this also appears in Aquinas's *Summa theologiae*, written several years before the commentary on *De caelo*, which links these suppositions explicitly to the theory of eccentrics and epicycles. It reads:

> Reasoning is employed in another way, not as furnishing a sufficient proof of a principle, but as showing how the remaining effects are in harmony with an already posited principle; as in astronomy the theory of eccentrics and epicycles is considered as established, because thereby the sensible appearances of the heavenly movements can be explained; not however as if this proof were sufficient, since some other theory might explain them.[79]

Some medieval astronomers, such as Bernard of Verdun, were not unwilling to attribute physical reality to eccentrics and epicycles,[80] and Aquinas seems here to be on guard against this type of mathematicism just as Albert was against its more metaphysical counterpart in the works of Grosseteste and Kilwardby. So he takes pains to elaborate the reasons why we cannot judge with certitude about the heavenly bodies,[81] and he shows no hesitation in acknowledging that Aristotle, in this matter, had mistaken a suppositional theory for established truth.[82] Yet, he observes, it is not a sign of stupidity or of presumptuousness to venture such explanations for facts that are refractory to human understanding, and, in fact, proposals from those who can discuss cases such as these "with greater certitude than the general run of men, . . . are more deserving of our gratitude than our censure." [83]

Aquinas was not the only medieval to make this distinction between a physical explanation that realistically accounts for the facts and a mathematical explanation that merely "saves the appearances." [84] Yet his endorsement of this way of looking at mathematical arguments in physical science, particularly since it was stated in the widely studied *Summa theologiae,* remained influential in Christendom all the way to the year 1615, when Cardinal Bellarmine wrote his famous letter to the Carmelite Foscarini stating that Galileo's attempts to justify Copernicanism are such that "all the celestial appearances are explained better than by the theory of eccentrics and epicycles." [85] Bellarmine, of course, was unwilling to accord a superior status to the Copernican theory, regarding it as just another attempt to save these same appearances.

3. *Experimental Method*

Thus far discussion has been limited to methodological considerations that relate to the use of observation and mathematics, and little has been said in a positive way about experimentation in the modern sense or the extent to which this may have been adumbrated in medieval science. Certainly Aquinas made no contributions in this direction. One of his contemporaries did make a start, however, and not long after his death a fellow Dominican completed what is perhaps the most outstanding piece of experimental work in the High Middle Ages. It is to these men, Peter of Maricourt and Theodoric of Freiberg respectively, that we must now turn our attention.

a. Peter of Maricourt

Little is known for certain about Peter Peregrinus of Maricourt, except that he wrote a much copied letter entitled *De magnete,* whose text we have, under date of August 8, 1269.[86] Circumstantial evidence and references to Peter contained in the writings of Roger Bacon enable us to surmise that he was born in the town of Maricourt, in Picardy, and that he made a pilgrimage (whence *peregrinus*) to the Holy Land, probably in connection with one of the Crusades.

Bacon refers to him as *magister Petrus* and *dominus experi-mentorum* and there seems little doubt from his terminology and philosophical concerns that he was university-trained, most probably at the University of Paris.[87] It was there, possibly, that Roger Bacon met him and became enamored of his experimental method. Bacon was most impressed by this extraordinary man and expressed his admiration in writing; his account is worth citing in its entirety.

> One man I know, and one only, who can be praised for his achievements in this science. Of discourses and battles of words he takes no heed: he follows the works of wisdom, and in these finds rest. What others strive to see dimly and blindly, like bats in twilight, he gazes at in the full light of day, because he is a master of experiment. Through experiment he gains knowledge of natural things, medical, chemical, indeed of everything in the heavens or earth. He is ashamed that things should be known to laymen, old women, soldiers, ploughmen, of which he is ignorant. Therefore he has looked closely into the doings of those who work in metals and minerals of all kinds; he knows everything relating to the art of war, the making of weapons, and the chase; he has looked closely into agriculture, mensuration, and farming work; he has even taken note of the remedies, lotcasting, and charms used by old women and by wizards and magicians, and of the deceptions and devices of conjurors, so that nothing which deserves inquiry should escape him, and that he may be able to expose the falsehoods of magicians. If philosophy is to be carried to its perfection and is to be handled with utility and certainty, his aid is indispensable. As for reward, he neither receives nor seeks it. If he frequented kings and princes, he would easily find those who would bestow on him honours and wealth. Or, if in Paris he would display the results of his researches, the whole world would follow him. But since either of these courses would hinder him from pursuing the great experiments in which he delights, he puts honour and wealth aside, knowing well that his wisdom would secure him wealth whenever he chose. For the last three years he has been working on the production of a mirror that shall produce combustion at a fixed distance;

a problem which the Latins have neither solved nor attempted, though books have been written upon the subject.[88]

The man revealed by an analysis of the *De magnete* agrees quite well with this description.[89] The letter was written to an intimate friend, Siger of Foucaucourt, while Peter was working as an engineer with the French army besieging the town of Lucera in southern Italy. Here he apparently conceived the idea of utilizing magnetic force to keep an astrolabe tracking the uniform motion of the heavens, and saw also the possibilities of constructing a perpetual motion machine on similar principles. To explain these devices to Siger, Peter found it necessary to review the principles of magnetism and the causes of magnetic phenomena, most of which he had discovered through his own ingenuity and experimentation. The astrolabe described in the letter no doubt inspired a passage in Bacon's *Opus maius* that has already been cited (*supra,* p. 50), and thus serves to confirm the identity of this man and his relationship to Roger Bacon.

The *De magnete* is divided into two parts, the first concerned with the study of the magnet in general and the second with the application of this knowledge to the construction of the mechanisms in which Peter was interested. The first part of the letter already shows Peter's skill as an experimenter, for here he details the procedures for locating the poles of a magnet and then uses these to demonstrate a series of magnetic properties. His method for locating the magnet's poles is the following. A loadstone of good quality should be procured, he writes, and this should be ground and polished until it is spherical; then an iron needle or filament is to be placed carefully on the stone and allowed to seek its own orientation, after which a line is to be drawn on the sphere marking what that orientation was.[90] Then the operation is to be done again with the needle or filament in another position, and the line corresponding to this also marked. "If you wish," writes Peter, "you may repeat this in many places or situations, and beyond doubt all the lines will converge at two poles, just as all the meridians of the world's sphere converge at its two opposite poles."[91] Naming the magnet's poles after the celestial poles, he decides to call one a north pole and the other a south pole. Another way of finding these poles, Peter observes, is to keep applying the point of the needle to different positions on the loadstone; the place where it adheres most frequently and firmly will be a pole. Yet a third method is to break off a small piece of the iron filament, about two finger-nail widths in length, and apply this to various positions on the spherical loadstone. If the filament stands perpendicular to the surface it is at a pole; if not, it should be moved until it does stand perpendicular and then a pole has been located.[92]

When the poles of the magnet are thus known, the loadstone should be mounted in a round wooden dish or container with both poles the same distance from its edge, and allowed to float in a larger container filled with water. If the apparatus is constructed properly the loadstone will turn the wooden dish in which it floats until its north pole points toward the north pole of the heavens and its south pole toward the south pole of the heavens. "And if you change this orientation a thousand times, the stone, at God's command (*nutu Dei*), will return just as often to its proper orientation." [93]

When the north and south poles of the magnet have thus been identified, they should be so marked for later reference. After two magnets have been marked, one can be mounted in a floating dish and the other brought close to it; then it will be found that the north pole of the one attracts the south pole of the other, and vice versa, while the north pole of the one repels the north pole of the other. Here Peter recognizes that he has discovered a *regula*, or law, which will later be referred to as the magnetic law of attraction and repulsion.[94] In the knowledge thus gained that unlike poles attract and like poles repel, Peter remarks that he hereby dispels "the foolishness of those who say that if scammony attracts red bile by reason of likeness, therefore a magnet will attract [another] magnet more than iron, which they falsely suppose, since the truth is what appears in the experiment." [95]

Next Peter explains how an oblong piece of iron may be magnetized, so that if it be made to float on a light piece of wood or on straw it will seek the north star. In such case the

part that touched the north pole of the loadstone will turn toward the southern part of the heavens. Peter takes pains to
observe here that the magnetized needle does not actually
point toward the north star, but rather toward the true pole
of the heavens.[96]

Other experiments with a magnetized needle disclose its
further properties when attracted or repelled by a loadstone.
The north pole of the stone will attract the south pole of the
needle, whereas it will tend to repel the needle's north pole.
If, however, the north pole of the needle be held firmly and
put in contact with the north pole of the stone, the poles of
the needle will be reversed and it will then seek the stone's
north pole rather than be repelled from it.[97] Peter also ascer-
tains that if a magnetized piece of iron, AD, where A is the
north pole and D the south, is broken in half to make two
magnets, AB and CD, the resulting polarities will be A north,
B south, and C north, D south, so that the two pieces will
naturally reunite in the orders ABCD or CDAB, but not in
the orders BACD and ABDC.[98]

Without going into further details on the instruments
constructed by Peter, from what has been said one can understand why his letter on the magnet was recognized by
later writers as a distinctive experimental contribution.[99] At
the same time, however, one should not fail to note the philosophical overtones of Peter's letter, e.g., his uses of the terminology of the Schools and the ordination of his experiments
ultimately to answering causal questions. Magnetism, like the
rise and fall of the tides, was regarded in Peter's time as an
occult work of nature; this in itself explains his concern to
make the causes of such phenomena manifest.[100] Peter also
approached his experimentation with certain presuppositions
in mind. For example, before explaining how to find the poles
of a spherical loadstone, he advises that "you ought to know
that this stone bears in itself a likeness of the heavens." [101]
Again, his instructions for mounting the loadstone when determining its poles are directed toward giving it freedom of
movement lest "the natural motion of the stone be
impeded." [102] Peter also searches for the causes of magnetic
induction, and explains the reversal of the poles of the lightly
magnetized needle as being caused by the fact that the "im-

pression of the last agent overcomes and alters the force of the first." [103] His experiments that involve breaking magnets into smaller pieces and reuniting them, finally, seem designed to prove the principle that "the agent not only intends to make its patient like unto itself, but to unite it in such a way that, from the agent and the patient a numerical unity will result." [104]

Peter employs the methodology of falsification with considerable skill when refuting those who hold that the loadstone derives its power from the place where it was mined and that it naturally tends to turn in that direction. [105] He himself is quite convinced that the magnet gets its power from the poles of the heavens, and this is why it aligns itself with them rather than point toward the north star. Peter would go even further, however, and say not only that the magnet's poles get their power from the celestial poles but that the magnet as a whole "receives its influence and power from the entire heavens." [106] To show this, he conceives an experiment wherein a spherical loadstone is pivoted at both poles and aligned in a north-south direction; the pivot of the north pole is then elevated at an angle so that the magnet's north-south axis points directly toward the celestial north pole. "Then," writes Peter, "if the stone rotates according to the movement of the heavens, you should rejoice that you have attained this marvelous secret; if not, however, the failure should be attributed to your own lack of skill rather than to nature's defect." [107]

This statement sounds suspiciously like one emanating from within the hermetic tradition, which was to surface later in the writings of Paracelsus, and it makes one pause over Roger Bacon's statement that Peter's work was really ordered to exposing "the falsehood of magicians." Although Roger refers to him as a mathematician, [108] moreover, there is no "metaphysics of light" apparent in his letter, and mathematical reasoning is not dominant, being similar where it is used to that employed by Aquinas in his proof of the earth's sphericity. Peter's statement that the "stone bears in itself a likeness of the heavens," his references to the magnet getting its power and influence from the heavens, and his assertions about its movements being natural and carried out at God's command, could be an indication that he subscribed to a type of Neoplatonism popular in the twelfth century. Here the universe was conceived as animated by a world soul (*anima mundi*) and the view of nature it sponsored was animistic, not unlike that later developed by William Gilbert.

b. Theodoric of Freiberg

Theodoric of Freiberg, unlike Peter Peregrinus, provides abundant material for study and enables us to draw a clearer picture of his contribution to experimental science.[109] His experimentation, like that of Peter, however, was done outside a university setting — a characteristic of experimental work that persisted until the seventeenth century — and probably would not have been recorded for posterity if Theodoric had not been a friar preacher. In 1304 he was present at the General Chapter of his Order, and while there he was told by his Master General, Aymeric de Plaisance, to write up his investigations on the rainbow. Theodoric thereupon wrote his treatise "On the Rainbow and Radiant Impressions" (*De iride et radialibus impressionibus*), a lengthy opus that runs to over 170 pages in its printed edition.[110] This extensive work was addressed to Aymeric, who ended his term as General in 1311; thus 1304 and 1311 are the terminal dates for the composition of the treatise, although it was undoubtedly based on work done previously, for otherwise Aymeric would not have known of Theodoric's adeptness in this science.

Theodoric studied at the University of Paris shortly after Aquinas's death and belonged to the same German Province (*Teutonia*) as Albert the Great, being one of Albert's successors as Provincial of that Province.[111] He wrote a large number of opuscula in philosophy and theology, and these serve to identify his general speculative orientation, which is closer to Albert's than to Aquinas's. Considerable eclecticism is manifest in his writings, although he is rather consistently Augustinian and Neoplatonist in his theology and Averroist and Aristotelian in his philosophy. Crombie has maintained an influence of Grosseteste on Theodoric from points of similarity in their optics;[112] howsoever this might be verified in particular details, Theodoric's Neoplatonism did not extend to his acceptance of a "metaphysics of light" nor to the mathematicist view of nature that this implied.[113]

Theodoric's place in the history of optics is guaranteed by his detailed analysis of the primary and secondary rainbows, of lunar and solar halos, and of other optical phenomena appearing in the earth's atmosphere. At a time when Peter Peregrinus provided the only real precedent for experimentation, Theodoric set about systematically investigating the paths of light rays that generate radiant colors in the earth's atmosphere, and did so largely by experimental means. He utilized spherical flasks filled with water, crystalline spheres, and prisms of various shapes to trace the refractions and reflections involved in the production of radiant colors. He also worked out a theory of elements that was related to his search for optical principles, and which stimulated experimentation along lines that could more properly be called verification than anything we have seen thus far.[114]

Theodoric's treatise, preserved in a number of manuscript copies whose scribes faithfully reproduced his many diagrams, has attracted the attention of historians of optics since the beginning of the nineteenth century. He thus has been hailed as a precursor of modern science and his work read as though he were using mathematical and experimental techniques developed only in the seventeenth century. There is no doubt that Theodoric's work foreshadowed Descartes' analysis of the rainbow and even may have provided a partial inspiration for this analysis, but at the same time its underlying methodology differs from the Cartesian. Theodoric's geometry derives from the optics of the Schoolmen and of Arabs such as Alhazen, whose *Perspectiva* he heavily utilized, and his angular measurements were based on medieval astronomy and the primitive trigonometry of Ptolemy's *Almagest*. But what is most characteristic of Theodoric's methodology is that, like the optical treatises of the medievals generally, it was situated in the framework of Aristotle's *Posterior Analytics*. Theodoric's research was directed toward ascertaining the causes of rainbows and of radiant phenomena, so that from these causes he could deduce all of their observable properties.

That this is Theodoric's intention is clear from his statement of methodology at the beginning of the treatise. He writes:

> The first thing to be considered is the statement of the Philosopher [Aristotle] in the *Posterior Analytics,* namely, that in the science of the rainbow it is the function of physics to determine the *quid* and of optics to determine the *propter quid.* The meaning of this statement, as is shown in the same book, is that definition is twofold, whether it be of the subject or of the property. One form of definition tells what the thing is in itself and absolutely, and this form, if it be of the subject, can be the principle of a demonstration, just as the subject itself to which this definition belongs is the principle of a demonstration; but if it be the definition of the property, the definition is the conclusion of a demonstration, just as the property itself is, according to the Philosopher in the same book. And according to the first mode of defining, it pertains to natural philosophy, in considering the rainbow, to determine its *quid,* as for example [the physicist] may tell in this or some similar manner that the rainbow is some sort of impression in rainy or cloudy air, of such and such quality as to the number, order, and position of colors, extended in an arc above the level of the horizon.[115]

Note in this account the erroneous reading of Aristotle, who states that it is for the physicist to determine the *quia* and the optician the *propter quid.* As we have already noted, however, in some manuscript copies of Grosseteste's *De iride* this same error is made, and it may be an indication of Theodoric's use of Grosseteste's opusculum. Whatever the source of his teaching, however, Theodoric apparently believes that the rainbow has a *quid* that can be known by the physicist, although the *propter quid* will pertain to geometrical optics. He goes on:

> There is another form of definition by which the being [*esse*] of a thing is determined, namely, by a definition telling both the *quid* and the *propter quid.* This is one and the same definition of the subject and of the property, giving the *quid* of the subject and the *propter quid* of the property, because it introduces the cause of the

property's being in the subject and is therefore the middle term of the demonstration; and, moreover, that definition and the whole demonstration differ only by position, according to the Philosopher. In this way it is the function of optics to determine the *quid* of the rainbow, because in so doing it indicates the *propter quid,* insofar as, to the aforesaid description of the rainbow, is added the manner in which this sort of impression is caused by luminous radiation going from some shining heavenly body to a determinate place in the cloud, and then by particular refractions and reflections of rays, it is directed from that place to the eye [of the observer].[116]

Here Theodoric, apparently puzzled by the text of the *Posterior Analytics,* justifies to himself that there is a way of understanding Aristotle to mean that geometrical optics can furnish not only the *quid* of the rainbow but also the *propter quid* of its properties. He then concludes his statement of methodology as follows:

For this reason, therefore, the science of optics subalternates to itself the science of the rainbow and of the other impressions produced by rays in the heavens. And because of what has been said it is convenient, indeed necessary, for optical and physical reasoning to be used together in the present matter.[117]

Faulty though his exegesis of Aristotle may be, Theodoric is convinced that his own work will be a mixture of physics and geometrical optics. As he appears to understand it, the proper subject of his study will be the atmospheric region of the heavens, with its clouds, mists, rains, and spherical droplets of various kinds, whose definitions pertain to physics, or more properly to meteorology as one of its branches. The properties he is investigating, on the other hand, are those produced by radiation passing through these droplets, and for this it is necessary to understand their spatial configurations and the geometry of the light rays passing through them, which involves him in a search for a mathematical type of *quid* and for the *propter quid* explanations of the properties that derive from this. Just what causes will be involved is not specified by Theodoric at this point, but these become apparent in his later analysis.

From this methodological introduction, it is to be expected that Theodoric's exposition will emulate the Aristotelian ideal of a demonstrative science, wherein the principles and causes are first established and then all of the properties of radiant phenomena are deduced therefrom. On superficial inspection this would seem to be very much an a priori approach to experimental science. As we have shown elsewhere, however, the structure of Theodoric's treatise is unintelligible if it is not understood in the context of the Aristotelian method of resolution and composition, a method already found in Grosseteste's work and brought to considerable perfection by the Averroist Aristotelians at Padua by the end of the sixteenth century.[118] The method was also explained and adapted by Albert the Great, who emphasized that any work of analysis or resolution, such as would be the demonstration of properties, must be preceded by a process of search and discovery, frequently employing dialectical methods, so as to determine the principles and causes to which the properties will be resolved.[119] Theodoric's treatment of the rainbow parallels, in fact, Albert's *via inventionis* and *via judicii;* the first part is concerned with a determination of the principles from which radiant colors are produced and with the material and efficient causes involved in their production, whereas the second part is concerned with a resolution of the rainbow's properties to these principles and causes. Experimental techniques are utilized in both parts, although more interestingly in the first, because it is here that the dialectical inquiry foreshadows the interplay between theory and experiment that characterizes modern methodologies.

The general setting of Theodoric's methodology is thus quite traditional, even though his use of it produced revolutionary results. Essentially, the novelty of Theodoric's contribution consists in the fact that he was not content simply to observe how rainbows are produced in nature, but rather attempted to duplicate the process under controlled conditions where he could observe in detail all of the component factors involved in the rainbow's production. Most of those who had searched for the material cause, or proper subject, of the rainbow's appearance thought that this must be a rain cloud. Even those who suspected that the individual raindrops were an important factor, as did Roger Bacon and Albert the Great, did not think of isolating the individual drop from the ensemble that produced the bow. Thus, when they compared the rainbow with the spectrum produced by the passage of light rays through a spherical flask filled with water, they thought of the flask as a miniature cloud or collection of raindrops. Theodoric's unique contribution was that he first saw "that a globe of water can be thought of, not as a diminutive spherical cloud, but as a magnified raindrop." [120] It was this discovery, together with the implicit realization that the entire rainbow is simply an aggregate of partial spectra produced by individual drops, that led him to the first correct explanation of the basic features of the primary and secondary rainbow. The underlying insight was what permitted him to immobilize the raindrop, in magnified form, and in what would later be referred to as a laboratory situation, where he could trace one by one the various factors involved in the rainbow's production.

Theodoric's work represents a remarkable achievement in geometrical optics, and yet an error in geometry vitiated in large part the quantitative aspects of his theory. The error consisted in using Aristotle's "meteorological sphere," at the center of which the observer is located and on whose periphery are found the sun and the raincloud. The advantage of using this sphere is that it enabled all calculations to be made by the methods of medieval astronomy, but it unfortunately committed Theodoric to a geometry wherein the sun and the raindrop would have to be thought of as at equal distances from the eye of the observer, and this of course was at variance with the facts. Theodoric also made an error either in measuring or in reporting the angle subtended by the arc of the primary rainbow at the eye of the observer, recording this as only 22°, whereas the correct figure is closer to the 42° registered by Roger Bacon.[121]

These defects in Theodoric's work notwithstanding, he was the first to trace correctly the various paths of light rays through raindrops to produce the primary and secondary rainbows, noting that the primary rainbow involves two refractions at the surface of the raindrop closer to the eye of the observer and one internal reflection at the farther surface, whereas the secondary rainbow also involves two refractions at the closer surface but requires two internal reflections at the surface farther from the observer.[122] Utilizing these paths of generation and his theory of color formation, Theodoric was able to explain that the rainbow always appears as an arc of a circle, that its colors always appear in a certain order, that each color is projected to the observer from a different raindrop or series of drops, and that an observer who changes his position sees a different rainbow because different drops are required for its appearance. With regard to the secondary rainbow, he was able to explain that its colors appear in an order the reverse of that seen in the primary bow, that its colors are less intense than those of the primary bow, and that it appears less frequently than the primary bow.[123]

Theodoric's aim, as we have seen, was to find the "true causes" (verae causae) of the generation of the colors of the rainbow,[124] but he never seemed completely satisfied with his "principles" for explaining radiant colors, and in fact wrote another opusculum, De coloribus, to clarify his thought on color in general.[125] It is in the portions of his treatise De iride, however, where he is trying to establish the principles of radiant color, that his methodology becomes most interesting. He discerns an analogy between a process described in his theory of the elements, worked out in two separate treatises entitled De elementis and De miscibilibus in mixto, wherein the four elements are generated from two sets of contrary qualities, and the process whereby four colors might be generated from two sets of optically contrary principles. In order to employ the analogy, however, Theodoric had first to establish that there were four colors in the rainbow, and this contrary to the teaching of Aristotle, who had maintained that there were only three. Theodoric's use of induction and a variety of experiments to disprove Aristotle and to establish his own principles dialectically comes closer to modern experimental method than any work that would appear before the seventeenth century.[126] Theodoric employed the term experiment (experimentum) at least twelve times in the De iride [127] and in three places explains that he has thought out experiments (experimento perpendimus) in order to establish his point.[128] Thus his is not a mere enumeration of experi-

ences, or a *universale experimentale* in the senses we have
seen in Grosseteste and Albert the Great, but a deliberate
empirical procedure, at times involving measurement, designed to verify or falsify a proposed explanation.

Other ways in which Theodoric adumbrated modern scientific methods may also be noted, but for our purposes it is
important that all of this was done in the context of an underlying Aristotelian methodology, aimed at discovering the
causes of the rainbow and deducing properties from these
causes by *propter quid* reasoning. Regarding his causal analysis, Theodoric states in one of his opuscula that it is for the
physicist to consider all four causes, final, efficient, formal,
and material, but in his own causal reduction he concentrates
mainly on the material and efficient causes of radiant phenomena, making only occasional mention of the form, and
none at all of final causality. The concentration on material
and efficient causality is easy to explain, since Theodoric regarded the rainbow and other radiant phenomena as accidents (i.e., as modifications of substances and not substances
themselves) of the atmospheric region, and he had treated of
the quiddity or *quid* of such phenomena in his opusculum *De
accidentibus*. There he teaches that the *quid* of an accident is
to be ascertained from its causes, and particularly its material
cause, or the proper subject in which it is found. Moreover,
since the rainbow is a radiant phenomenon and thus produced by the agency of light, the mode of action of light as
an efficient cause in its production will be most helpful in understanding the form, or appearance, that is finally generated.
This is the reason why Theodoric is intent on finding the material and efficient causes of radiant phenomena, for such
knowledge reveals their *quid*, insofar as such entities can be
said to have a *quid*. Viewed from another aspect, and this is
pointed out in the methodological introduction already
quoted, the radical subject of Theodoric's investigations is not
so much the rainbow or radiant phenomena as the atmospheric region of the heavens, of which the rainbow turns out
to be an accidental property. Thus his possibility for achieving *propter quid* demonstration is based on the fact that he
has sufficient knowledge of the *quid* of the atmospheric region in the order of substance, i.e., in terms of the clouds,

mists, and raindrops that go to make it up, that he can give a *propter quid* explanation of the appearance of certain properties, in the order of accident, that inhere in the atmospheric region as a subject. In this context, Theodoric's failure to mention final causality is perfectly consistent with the technique of demonstration *ex suppositione finis* already discussed in connection with Aquinas's methodology. In this procedure, the form to be generated, i.e., the appearance of the rainbow, is itself the end to be attained in the process of generation, and therefore it is taken as the final cause *ex suppositione,* but as such dictating the necessity of all the other causes. Thus, *if* the rainbow is to be generated, then such-and-such an efficient cause must operate on such-and-such a matter in order to produce this particular form. Since, therefore, the final cause, or the form intended and attained in the process of the rainbow's generation, is the starting point of the methodological process by which the other causes are demonstrated, the resolution to the final cause does not appear in the analytic process.[129]

Through all of this methodological development in the writings of Peter and Theodoric, there is an underlying philosophical realism analogous to that found among the pioneers in optics at Oxford University. Theodoric had seen enough of the errors of his predecessors in their attempts to unravel the mysteries of the rainbow; he knew that much of their work was mere conjecture concerning the rainbow's causes, and so he set himself resolutely to discovering its "true causes" by a strict demonstrative process. Precisely because of his opposition to Aristotle's three-color theory, moreover, Theodoric felt compelled to justify the reality of a fourth color when he wished to use it in his own explanatory scheme. This led him to raise the question whether radiant colors are real, and if so where and how they exist, despite Roger Bacon's earlier dismissal of such colors as merely an appearance produced by defective human vision.[130] Albert the Great's empiricism thus had its effect on Theodoric and his work. The Paris School owed some of its inspiration to Oxford, but its search for causes was less markedly in the mathematical sphere and more in the real, physical world. Such a concern would continue to manifest itself at Paris as the fourteenth century continued on, and particularly as the problems of motion came to be newly addressed there in light of the approaches inaugurated at Oxford by Ockham and the Mertonians.

4. *The Paris Terminists*

Thus far we have presented evidence for a more realist orientation at the University of Paris than at Oxford University deriving from an empiricist, as opposed to a mathematicist, approach to nature. Another influence bearing on the issue of realism, of equal if not of greater importance than the empiricist approach, is that deriving from the condemnations of 1270 and 1277, which had a wider and longer-lasting effect at Paris than any of the ecclesiastical censures at Oxford.[131] When Étienne Tempier condemned the proposition, for example, that "God could not move the heavens with rectilinear motion because a vacuum would remain," [132] he not only denied an accepted Aristotelian thesis but he opened the way for a new view of reality that could be seen by the eyes of faith. Pierre Duhem has made much of these condemnations and their role in forging the beginnings of modern science. Essentially his thesis is that science was thereby freed from the shackles of adherence to Aristotle's *Physics* and that man's imagination was given the incentive to construct all types of hypothetical schemata for "saving the phenomena." [133] Duhem's instincts were right, even though the interpretation he placed on the condemnations is not completely correct. Rather than urge the Paris terminists in the direction of imaginary mathematical constructions, a path already well trod by the Mertonians, Tempier's action committed them to the belief that reality was not exhausted by Aristotle's view of nature, but that God's "absolute power" could be used as a valid argument to explore new facets of the real world. The result was that the Paris terminists, following the lead of Jean Buridan, subscribed to a realist view of local motion, and consistent with this, sought the causes and forces that might exist in nature and that could account for such motion. This led them to make outstanding contributions in dynamics, whereas the Oxford contribution was essentially in kinematics, and even to suggest physical explanations for the motions of the heavens in terms of universal dynamical principles. In the light of such contributions by Buridan and his disciples, Albert of Saxony and Nicole Oresme, one can understand why Duhem was tempted to label these fourteenth-century Parisians the true "precursors of Galileo."

a. Jean Buridan

The role of Jean Buridan at Paris was not unlike that of Thomas Bradwardine at Merton College.[134] Buridan does not seem to have been directly influenced by Ockham, but he had a great interest in formal logic, particularly the terminist logic of such authors as William of Sherwood and Peter of Spain, and he made remarkable contributions to both the theory of supposition and that of consequences. In material logic he subscribed to the Aristotelian concept of science, however, and understood well its theory of demonstration. In fact, his understanding of how the natural sciences demonstrate *ex suppositione* and his entire treatment of final causality in nature builds on the discussion of these topics already seen in Aquinas, and is one of the most balanced accounts in the fourteenth century.[135] By not claiming too much for the natural sciences, Buridan could insist on empirical procedures proper to them and not confuse the resulting necessity with that of either logic or metaphysics. His was not a skeptical view, moreover, since he believed that one could attain to knowledge of causes and through these could have certitude in the physical sciences. There is even some evidence that while Buridan was rector of the University of Paris he censured the teachings of Nicholas of Autrecourt, who had skeptical views of causal inference and on this account questioned the possibility of a science of nature that would be based on causal laws established by inductive generalizations.[136]

It is in his theory of motion, however, that Buridan's most distinctive contributions to the rise of modern science were made.[137] Although definitely nominalist in his logical sympathies, Buridan reacted against the Ockhamist conception of motion and himself fostered such a realist view of this phenomenon that he perforce had to be concerned with its causes and effects. In working out this position, Buridan distinguished between local motion and all other kinds of motion; he was willing to concede, in this connection, that in qualitative change the terminus attained is a quality that inheres in the subject, and therefore that the motion involved in alteration is essentially a *forma fluens* that could be identified with the motion's terminus.[138] In local motion, on the other hand, since this involves change of place, where the terminus of the motion is not something inhering in the moving object but is rather an external determination of it, there is no possibility of the motion's being identified with the terminus it attains.[139] Again, for Buridan local motion was itself a reality, independent of place. To establish this, he simply referred to the famous *articulos Parisienses* and concentrated on the example therein provided of the motion of the outermost heavenly sphere.[140] Here change of place could not be spoken of, since the outermost sphere is not in place, and yet the condemnation of 1277 had forbidden anyone to maintain that this sphere could not be set in rectilinear motion by the absolute power of God. Therefore motion must be something real and independent of both the thing moved and the place that ordinarily terminates its movement. Thus for Buridan, as opposed to Ockham, motion is more than a mere word; rather it is something real, a *res pure successiva,* or in Burley's terminology, a *fluxus,* and as such it cannot be located in a category. Anneliese Maier reads the following consequences into Buridan's conception, namely, that for him motion "is not further ontologically analyzable; it is given as real, empirically discovered as an instantaneous state of the thing moved, but different both from this and the place in which it is; and it cannot be further clarified nor does it require clarification." [141]

Since Buridan was thus convinced that local motion, as exemplified in the case of the thrown object, was really a "new effect," to use Ockham's expression, it is not surprising that he concerned himself with the cause of such motion. Here, like many commentators who had studied Aristotle's treatment of the projectile problem, Buridan was dissatisfied with the causal explanation given in the *Physics* and also with the various emendations of this doctrine current in his time. It is probable that he was here influenced by Franciscus de Marchia, a Franciscan theologian who taught at Paris around 1322. Like Duns Scotus, whose teachings he followed, Franciscus was convinced of the reality of local motion; he explained the continuing motion of the projectile by a *virtus derelicta,* or "force left behind" by the projector after initial contact with the projectile had been broken.[142] A similar force, according to Franciscus, was also left behind in the medium to assist the motion, and here his teaching was consonant with Aristotle's original explanation. The *virtus* or impressed force, in Franciscus's view, was of the temporary, self-expending kind, and he proposed it as the cause of projectile motion on the strength of a principle of economy (*quia frustra fit per plura quod potest fieri per pauciora*) and because he thought that through its use all the appearances are better and easier accounted for (*melius et facilius salvantur omnia apparentia*).[143] Finally, it occurred to him to apply this theory even to the heavenly bodies, for he states that "with the intelligences ceasing to move the heavens, the heavens would still be moved, or revolve, for a time by means of a force of this kind following and continuing the circular motion . . ."[144]

Buridan's teaching differed from Franciscus's in that, whereas for the latter the *virtus derelicta* was a temporary impressed force, for Buridan it became a permanent quality, which he called impetus and even quantified in a general way in terms of the primary matter of the projectile and the velocity imparted to it.[145] As a permanent quality it would not be self-expending and thus would have to be overcome by contrary resistances; otherwise it would last indefinitely (*ad infinitum*).[146] Like Franciscus, Buridan used his concept to explain the motion of the heavens, dispensing with intelligences and holding that God imparted an impetus to the heavenly bodies at the time of the world's creation. He made no distinction, again, between rectilinear and curvilinear impetus, regarding the quality as capable of sustaining either straight-line or circular motion. Buridan also applied his theory to the problem of the falling body, explaining that the continued acceleration of the body results because its gravity continually impresses more and more impetus.[147]

Buridan's justification for his theory of impetus was more empirical than Franciscus de Marchia's; in fact, he uses the terms *experimentum* and *experientiae* in ways quite similar to the tradition we have already discussed.[148] For example, he introduces his treatment of the curvilinear impetus in a smith's wheel with the words "*et experimentum habetis,*" and then discusses how hard it is to bring a rotating stone suddenly to rest, but how the impetus will diminish from the resistance arising from the stone's gravity, and also how, if there were no such resistance, the stone would keep revolving perpetually.[149] In proposing his theory of rectilinear impetus, Buridan details a whole series of *experientiae,* which in effect are quasi-experimental observations of objects in inertial motion, in order to disprove the standard Aristotelian explanations and support his own theory.[150] Again, as Clagett has pointed out, when Buridan is refuting the various explanations of the cause of acceleration of falling bodies that were current in his day, his refutation has a "strongly empirical and observational character"; here he uses, among other things, the force of impact as a practical measure of the velocity, much as Leonardo da Vinci and Galileo were later to do.[151]

Like Franciscus de Marchia, Buridan invoked hypothetical argument to support his positions, showing that they could not be disproved as could alternative proposals and that all the appearances were in harmony (*omnia apparentia consonant*) with his explanation.[152] Duhem has claimed this as yet another instance of "saving the appearances,"[153] and others have noted similar cases of inconclusive reasoning on Buridan's part. For example, after proposing that God could move the heavenly bodies by an impressed impetus, he adds: "But this I do not say assertively, but [rather tentatively] so that I might seek from the theological masters what they might teach me in these matters as to how these things take place."[154] Again, when discussing Aristotle's rules involving the ratios of motions, Buridan raises the question whether these are "universally true."[155] In answer he observes that some of the rules require constant forces and resistances, and then he goes on:

> And from these things it seems to me that it must be inferred that these rules are rarely, or never, found to produce their effect. Nonetheless, these rules are conditional

and true, for if the conditions stated in the rules were observed, everything would occur just as the rules assert. For this reason it ought not to be said that the rules are useless and fictitious because, although these conditions are not fulfilled by natural powers, it is nevertheless possible, in an absolute sense, for them to be fulfilled by the divine power.[156]

It has been observed of this that it is a "purely hypothetical discussion," [157] but surely the sense of "hypothetical" here is different from that used to characterize a theory such as that of eccentrics and epicycles. In fact, contrary to Duhem, it is precisely Buridan's commitment to realism and to a realist theory of motion that enables him to maintain that the rules are true, even though they might be conditional. Here his reasoning is very similar to that of Aquinas in his treatment of demonstration *ex suppositione,* which Buridan himself had justified as the proper methodological procedure in natural science.[158]

In conjunction with the theory of eccentrics and epicycles, it is of interest that Buridan raises the question whether one can prove that the earth is at rest and that the heavenly spheres rotate around it, as opposed to the heavenly spheres' being at rest and the earth's rotating on its axis once a day. With regard to this second possibility, Buridan voices the doubt whether "all the phenomena that are apparent to us can be saved" (*possent salvari omnia nobis apparentia*) in this way.[159] He then discusses the relativity of motion and concedes that the heavenly appearances can indeed be saved by the second alternative; in fact, from the viewpoint of economy of explanation, this even appears to be the more reasonable explanation:

> Just as it is better to save the appearances through fewer causes than through many, if this is possible, so it is better to save [them] by an easier way than by one more difficult. Now it is easier to move a small thing than a large one. Hence it is better to say that the earth, which is very small, is moved most swiftly and the highest sphere is at rest than to say the opposite.[160]

"But still," continues Buridan, "this opinion is not to be followed." [161] He thereupon enumerates the various appear-

ances [*apparentiae*] that should be manifest on the earth's
surface to observers if the earth were rotating rapidly, and
yet which are not observed. The last of these, "which Aristotle notes is more demonstrative in the question at hand," is
that an arrow shot straight upward from the earth's surface
will fall back to the very spot from which it departed.[162] Buridan admits that those who advocate the earth's diurnal rotation "respond that 'authority does not demonstrate' and that
it suffices astronomers that they posit a method by which appearances are saved, whether or not it is so in actuality." [163]
But his own reserve, it would seem, arises from the fact that
he is not content with merely saving the appearances, but favors the explanation that for him best accords with reality.[164]

b. Albert of Saxony

Buridan was very much the natural philosopher and the
logician, and the mathematical mode of argumentation perfected by the Mertonians found little place in his work. His
disciple, Albert of Saxony, however, had a greater interest in
mathematics and incorporated the findings of Bradwardine
and his followers into his own expositions. Albert was more
"a transmitter and an intelligent compiler of scientific ideas"
than an original investigator, writes E. A. Moody, and yet, he
continues:

> Despite his lack of originality Albert contributed many
> intelligent discussions of aspects of the problems dealt
> with, and he had the particular merit of seeing the im-
> portance of bringing together the mathematical treat-
> ments of motion in its kinematic aspect, stemming from
> the Oxford tradition of Bradwardine, with the dynamical
> theories that Buridan had developed without sufficient
> concern for their mathematical formulation.[165]

Since Albert's works appeared in many editions soon after the
invention of printing, and hence were widely diffused, he in
fact became one of the principal sources of the knowledge of
late medieval science in the Renaissance.

On the reality of motion Albert subscribed to basically
the same position as Buridan, although this is not easy to discern from his Questions on the *Physics* of Aristotle.[166] Here
he presents a somewhat ambivalent treatment, possibly re-

flecting his greater concern with the nominalist arguments deriving from William of Ockham. Albert devotes two questions to the problem of how local motion may be said to differ both from the moving object and the terminus it attains. The first, Question 6 of Book 3, asks "Whether anything that might be a certain *fluxus*, distinct from both the moving object and [its] place, is required for something to move locally?" [167] Thereupon he lists thirteen arguments on the affirmative side, followed by seven arguments on the negative side, and then gives his own reply. This is that such a *fluxus* is not necessary, arguing, among other reasons, from the analogy that it is not required in alteration. Albert thus concludes this question as a nominalist, and responds to all thirteen arguments, resolving their difficulties along nominalist lines. Immediately following this, however, in Question 7 of Book 3, he repeats the question with a qualification, viz., "Whether, admitting divine cases (*casus divinos*), one would have to concede that local motion is a thing distinct from the object moved and from [its] place?" [168] This time he answers the question in the affirmative, first by listing the same seven arguments he had given on the negative side in the previous question, then giving his reply as a realist, invoking the "articulum Parisius," and finally responding to the seven arguments, now resolving their difficulties along realist lines. Depending on which question one would read, one would get the impression that Albert was a nominalist or a realist. Actually, he seems to be both: following logical and natural reasoning, he sides with the nominalists; but "according to truth and the faith," he sides with the realists.[169]

Albert's presentation of the impetus doctrine of Buridan also resulted in modifications; in some respects these represent a regression from Buridan's work, since Albert effectively blurred the distinction between a self-expending *virtus impressa* and a permanent inertial impetus. Also, in analyzing the case of a projectile thrown upward, Albert introduced a period of rest between the completion of the projectile's ascent and the beginning of its descent. When he applied this theory to the case of the projectile shot horizontally from a cannon, Albert speculated incorrectly about the resulting trajectory; despite this, his discussion proved fruitful for later

writers such as Leonardo da Vinci.[170] Of similar interest is
Albert's account of the cause of the acceleration of falling
bodies and his attempts to describe such acceleration mathe-
matically. He discusses various functional relationships between velocity and both distance and time of fall; none of
these provided the correct law of falling bodies, and yet they
stimulated later writers to continue work on this important
problem.[171] Again, Albert raised the question of the earth's
rotation and came to a conclusion similar to Buridan's. In
fact, he attempted to analyze the example of the arrow's
being shot upward in terms of Buridan's impetus theory, but
maintained, in Moody's words, "that the lateral impetus
shared by the projectile with that of the surface of the rotating earth would be insufficient to carry it over the greater arc
which it would have to traverse, when projected outward
from the earth's surface, in order to fall back at the same
spot." [172]

c. Nicole Oresme

Far more original than Albert, and indeed one of the
most creative minds in the fourteenth century, was his contemporary and fellow-disciple of Buridan, Nicole Oresme.
Oresme was a theologian as well as a master of arts, and he also
functioned effectively as a court adviser, preacher, and translator of scientific works from Latin into the French
vernacular. But his true skill lay in his mathematical ability
and in his imaginative treatment of physical and cosmologi-
cal problems, many of which anticipate developments that
were to be forthcoming only in later centuries. Unfortunately
the great bulk of his scientific writings were not published,
although they did exist in manuscript copies and thus exerted
their influence, if somewhat selectively, on the learned world.

One of Oresme's more frequently cited contributions is
the analogy he draws between the workings of a clock and
the universe. Thus he writes:

> When God created the heavens, He put into them motive
> qualities and powers just as He put weight and resis-
> tance against these motive powers in earthly things.
> These powers and resistances are different in nature and
> in substance from any sensible thing or quality here

> below. The powers against the resistances are moderated
> in such a way, so tempered, and so harmonized that the
> movements are made without violence; thus, violence ex-
> cepted, the situation is much like that of a man making a
> clock and letting it run and continue its own motion by
> itself.[173]

Although Oresme conceived this mechanistic type of explanation, it should be noted that he did not on this account dispense with intelligences as movers of the heavens, and he
continued to stress the essential differences between the mechanical principles relating to terrestrial motions and those
that governed the movements of the heavens. Again, it is
noteworthy that Oresme himself departed from Buridan's concept of a permanent impetus, regarding it as self-expending
from the very fact that it produces motion, and leaving some
confusion as to whether the impetus tends to produce a uniform motion or an acceleration. Since he seems to have favored the latter view, there is some difficulty in understanding how he envisaged impetus as an explanation of the
heavenly motions, which he of course insisted were uniform.[174]

   Typical of Oresme's sophisticated mathematical contributions was his attempted demonstration of the incommensurability of celestial motions, which he regarded as an important
argument for rendering the practice of astrology fallacious in
principle. This is found in a lengthy treatise, *De proportionibus proportionum* (On the ratios of ratios), which takes off
from Bradwardine's *Tractatus de proportionibus* and investi-
gates rather thoroughly the problem of relating ratios expo-
nentially.[175] It is in this context that Oresme makes a distinction between irrational ratios whose fractional exponents are
rational and those whose exponents are themselves irrational.[176] Such a distinction was a great mathematical advance for his day, and from it Oresme was able to deduce the
probable conclusion that the ratio of any two unknown ratios
relating to the movement of the heavens would be irrational.
Since he conceived of astrological prediction as based on the
exact determination of the conjunctions and oppositions of
heavenly bodies, which would not be calculable if irrational

ratios were involved, it became impossible in the light of this discovery to put astrology on a scientific basis.[177] Also noteworthy in Oresme's treatment of astronomical problems is his lengthy gloss on Aristotle's discussion of the possibility of a plurality of worlds, where he speculates on the existence of an infinite void space beyond the finite cosmos.[178]

Like Buridan and Albert of Saxony, Oresme addressed the problem of the possible rotation of the earth, and stressed with them the complete relativity of the detection of any motion from visual appearances, with the result that celestial phenomena would be just as well saved by the earth's diurnal revolution as by that of the heavens. Yet, his own conclusion he expresses in these terms: "The truth is, that the earth is not so moved but rather the heavens," and then he goes on to add, "however, I say that [this] conclusion cannot be demonstrated but only argued by persuasion." [179] His conviction in this case clearly did not derive from scientific demonstration; rather it came from faith, which he, like the other Paris terminists, regarded as a reliable source of knowledge concerning the physical universe. And although he did not believe that the earth rotated, he was willing to grant the possibility of the earth's moving with a small motion of translation, brought about by the fact that its center of gravity was being altered constantly by geographic changes and would always tend toward alignment with the center of the universe.[180]

Oresme also had some original contributions to make to the study of falling motion. He came the closest of all the Parisians to formulating a correct law of falling bodies, stating that the velocity is proportional to the time of fall, although he did not apply the Merton mean-speed theorem to calculate the distance that would be traversed. He knew the mean-speed theorem, of course, and in fact gave the first geometrical demonstration of the theorem through the use of his configurational geometry. Oresme accounted for the acceleration of a falling body in terms of an impetus acquired during the fall, and even conceived of the case where the earth might be pierced through its center along the path of fall. In this event the falling body would continue past the center of gravity until the impetus it had acquired was dissipated; then it

would fall back again toward the center and continue in os-
cillations of decreasing amplitude until it ultimately came to
rest.[181]

A final contribution of Oresme, and one of fundamental
importance in the development of modern science, is his treatise on configurational geometry entitled *Tractatus de configurationibus qualitatum et motuum*.[182] The work anticipates
Descartes' analytical geometry in several important respects,
particularly in the way in which it utilizes two-dimensional
figures to represent geometrically variations in the intensities
of qualities and in the velocities of motions with respect to
space and time respectively. The term *configuratio*, in
Oresme's teaching, actually has two meanings.[183] In its first im-
position it applies to the imaginative use of geometrical figures to graph the distribution of intensities, say, of a quality
in a body, and then, in a second and derived sense, it applies
to some kind of internal arrangement of intensities thought to
be inside the body and to characterize it essentially. It is significant that Oresme was not content merely to plot the configurations of qualitative and velocity variations, as in the
first usage, but wished also to use them in the second way to
explain sonic, musical, psychological, and even magical effects. Thus he conceived his mathematics as a way of uncovering the deeper secrets of nature through a knowledge of the
underlying causes that produce them. "Perhaps by proceeding in this way," he concludes a treatment of the difformity of
velocities, "there can be assigned causes of certain effects the
reasons for which are otherwise unknown." [184]

5. *The Paris Contribution*

With this we conclude our survey of the development of science to the end of the fourteenth century. The possibility expressed by Oresme in the very last citation is about the closest one would come, at Paris, to a suspicion of an underlying
mathematical structure that might be discovered through observation and yield true causal knowledge. Here there is an
echo of Grosseteste's thought, but with the difference that
Oresme was not too sanguine, as we know from his other writings, about the possibility of nature's fitting a rational mathe-

matical scheme. This is not to say that the work done at the University of Paris, or under its influence, was unappreciative of the value of mathematics in physical science. From Albert the Great and Aquinas to Albert of Saxony and Oresme there was constantly manifest an openness to the employment of mathematical argument in physics, whether such employment was to yield a *propter quid* demonstration of some quantitative modality or was merely to "save the appearances."

What is distinctive of the Paris contribution as opposed to Oxford's, however, was that at Paris there was no facile expectation that mathematics would yield a privileged insight into the world of nature. Because of this, the empirical temper of the purer Aristotelian tradition was more in evidence and bore fruit in the works of such men as Albert the Great, Peter of Maricourt, Theodoric of Freiberg, and the Paris terminists. Such an empirical attitude was based on a confidence in man's ability both to know the world of nature and to discover the causes of its phenomena. Also more in evidence at Paris was the effect of ecclesiastical censures, which led the terminists particularly to the conviction that faith could solve problems which might remain unresolved if viewed in the light of reason alone. It was this conviction that led them to regard local motion as real and as an effect that would require its own proper causes. Without this commitment they might never have developed their theories of gravity and impetus, or studied the various forces and resistances that affect motions in both the terrestrial and celestial regions. With it, they could make a good start on dynamical problems, and when the kinematical contributions of the Mertonians became available, incorporate them into the beginnings of a systematic science of mechanics.

By the end of the fourteenth century, therefore, mathematical ways of reasoning in natural science had become fairly sophisticated owing to the work of Grosseteste and his school, the Mertonians, and Paris terminists such as Oresme,[185] while a beginning had also been made in empirical and experimental approaches, based on a realist philosophy, owing to the efforts of such thinkers as Albert the Great, Aquinas, Theodoric of Freiberg, and Jean Buridan. What was required now was a more explicit concern with experimentation, and this particularly as ordered to the verification of laws and theories expressed in mathematical terms. For such a development we shall have to move to Italy, to study there some of the distinctive contributions at the University of Padua, where, over the next two centuries, the groundwork would gradually be laid for the work of Galileo.

CHAPTER FOUR

Padua and the Renaissance

THE principal cause of the condemnations of 1270 at Paris and of 1277 at Paris and Oxford was the movement known as Latin Averroism. The commentaries of Averroës on Aristotle were of inestimable help in understanding "the master of all who know," and at first they were used with enthusiasm by Christian scholars, particularly at the University of Paris. A detailed study of Averroës's writings, however, soon revealed teachings that were at odds with Christian revelation and that gave rise, on this account, to a heterodox Aristotelianism. Albert the Great and Thomas Aquinas were suspect of such heterodoxy in certain quarters, particularly among the Franciscan followers of St. Augustine. Much more heterodox, however, and vigorously opposed on this account even by Albert and Aquinas, was the Latin Averroism of Siger of Brabant and Boethius of Dacia. This exerted great influence in the arts faculty at Paris in the 1260's and 1270's until the movement was officially suppressed in 1277, when both the heterodox Aristotelianism of Siger and the orthodox Aristotelianism of Aquinas came under attack. The aftermath of the condemnations saw Aristotle somewhat discredited, and the way prepared for nominalism and for the more Neoplatonist type of scholasticism associated with Avicenna and advanced by Henry of Ghent and Duns Scotus, among others.[1] Latin Averroism thus effectively died at Paris, although it had some fourteenth-century proponents there, such as John of Jandun. But like most strong philosophical movements, it did not die completely, but rather re-emerged in a different setting and took on a new life. The new setting was in northern Italy, particularly at the University of Padua, which, along with Oxford and Paris, has been singled out by scholars as the birthplace of modern science.[2]

The movements we have studied thus far have pertained to the thirteenth and fourteenth centuries; that at Padua, on the other hand, started only feebly in the fourteenth century and progressed rather strongly throughout the fifteenth and sixteenth centuries. Some of the work of the fifteenth-century Paduan calculators, moreover, stimulated interest outside Italy, particularly in France and in Spain, and thus contributed to the growth of a broader movement that reached its culmination during the late sixteenth century in the work of Galileo at Padua. For the purposes of our study, which has been concentrating on the role of the *Posterior Analytics* in the development of scientific methodology, and particularly the various factors relating to the mathematical and experimental aspects of that methodology, it will be convenient to divide this complex development into three phases. The first will treat of the rise of the School of Padua and will examine in some detail the works of Paul of Venice and Gaetano da Thiene and those they influenced; the second will concentrate on developments in France and Spain that were at least partially inspired by these scholars; and the third will again focus on Padua to examine the contributions of Agostino Nifo, Jacopo Zabarella, and others who brought the scientific tradition at that university to its culmination.

1. Early Paduans

The concern with methodology that was to characterize the University of Padua in the late Middle Ages is seen in one of its early doctors, Pietro d'Abano, frequently called "the Conciliator" because of his major work, *Conciliator differentiarum philosophorum et praecipue medicorum*. Pietro is regarded by some as the source of the Averroism that later came to dominate at Padua, although this is contested by

others; his interest in natural philosophy and medicine, however, and the harmony he attempted to promote between the
two, had a lasting effect in the School of Padua.[3] Pietro understood both the *Posterior Analytics* of Aristotle and the
medical procedures laid down by Galen, and he unified them
precisely by insisting on the proper orders of resolution and
composition that we have already seen mentioned in Grosseteste.[4] Thus, in attempting to answer the question whether
medicine is a science, Pietro first points out the distinction
between sciences *propter quid* and *quia,* and writes:

> Science in the most proper sense is that which infers the
> conclusion through causes which are proximate and im-
> mediate, like that science defined in the *Posterior Analyt-
> ics,* Bk. 1, chap. 2 . . . And this kind of science is gained
> from demonstration *propter quid,* or what Galen called
> *doctrina compositiva.* There is a second sense of science
> that is also proper, and indeed can be said to be for us
> most proper; since for us the natural way is to proceed
> from what is more knowable and certain for us to what is
> more knowable in the order of nature: see the beginning
> of the *Physics.* When, in cases where effects inhere in
> their causes according to an essential order of priority,
> we arrive by the opposite order at the cause we are seek-
> ing, through proximate and logically immediate middle
> terms; or when we conclude an effect from more general
> causes, omitting certain intermediate causes, we acquire
> knowledge by demonstration *quia,* or what is called *doc-
> trina resolutiva.*[5]

Pietro's identification of *propter quid* demonstration with a
compositive procedure is based on his awareness that if one
knows the cause of a phenomenon, he can compose these in
his mind so as to explain the effect, and thus he has explanation or proof through the cause, or *propter quid.* If, on the
other hand, he does not know the cause, by a process of discovery he may be able to resolve the phenomenon back to it,
and then he will be able to go from effect to cause, and his
demonstration will be *quia.* Later on in the *Conciliator* Pietro
amplifies this teaching when discussing the prologue of Galen's *Tegni,* or art of medicine, where Galen has outlined three
doctrinae or ways of teaching medical science, i.e., by resolution, by composition, and by definition. Galen wrote:

In all the ways of teaching (*doctrinae*) which follow a definite order there are three orders of procedure. One of them is that which follows the way of conversion and resolution (*dissolutio*); in it you set up in your mind the thing at which you are aiming, and of which you are seeking a scientific knowledge, as the end to be satisfied. Then you examine what lies nearest to it, and nearest to that without which the thing cannot exist; nor are you finished till you arrive at the principle which satisfies it . . . The second follows the way of composition, and is the contrary of the first way. In it you begin with the thing at which you have arrived by the way of resolution, and then return to the very things resolved, and put them together again (*compone eas*) in their proper order, until you arrive at the last of them. . . . And the third follows the way of analysing the definition.[6]

What is important about this methodological beginning, as Randall has pointed out, is that it effectively transformed what had been in the minds of many merely a method of confirmation or proof into a method of discovery.[7] Other investigators, as we have seen in the case of Theodoric of Freiberg, located their scientific discoveries in the context of the *Posterior Analytics,* but they did not consciously elaborate a logic of discovery that might assist others to explore systematically the secrets of nature. The Paduans, on the other hand, maintained a consistent interest in methodological questions of this sort, and particularly developed the method of resolution to the point where it became, in the late sixteenth century, an effective tool of scientific discovery.

The similarity between Pietro's concept of method and that already pointed out by Averroës can be seen from the account given by Urban the Averroist in a commentary on the *Physics* written in 1334. Here he points out that Averroës has indicated three methods of demonstration, of which the first proceeds from causes that are more known both in the order of nature and to us and thus provides the greatest certitude, to be found in the mathematical sciences. A second method is that which proceeds from effects more known to us, and which lead us to the knowledge of causes that are more known in the order of nature; this second type of demonstration is referred to as a "sign" and does not have the force of the first type. A third method is that of

> demonstrations which proceed from causes which, though they are always prior and more known in the order of nature, are often posterior and less known to us. This occurs in natural science, in which from those things prior for us, such as effects, we investigate their causes, which are posterior and less known to us. And this is the way of the method of resolution. But after we have investigated the causes, we demonstrate the effects through those causes; and this is the way of the method of composition. Thus physical demonstrations follow after mathematical demonstrations in certainty, because they are the most certain after those in mathematics.[8]

With the teaching of Aristotle thus not only in harmony with, but actually reinforced by, that of Galen and Averroës, the way was prepared for a general acceptance of this methodological doctrine and its development by later scholars.

a. Paul of Venice

Perhaps the most important figure in the School of Padua was Paul of Venice, an Augustinian friar who studied at Oxford as well as at Padua, and who taught at the University of Paris and also at a number of universities in northern Italy. Paul not only continued the methodological interests of Pietro and Urban, but he imported into Italy the logic and the calculatory techniques he had learned at Oxford and at Paris, and combined these with an Averroist type of realism that gave a distinctive orientation to his thought and to that of his disciples. For this reason he may be regarded as the founder of the Paduan School, somewhat after the fashion of Bradwardine at Merton College and Buridan at the University of Paris.[9]

Apart from his numerous writings on nominalist logic, Paul composed a commentary on the *Posterior Analytics* that shows a detailed knowledge of the commentaries by Averroës, Grosseteste, Aquinas, and Giles of Rome—the last of whom, being the first Augustinian master at the University of Paris, receives special treatment at his hands.[10] Most of the examples we have already discussed are analyzed by Paul

with both perceptiveness and fidelity to the Aristotelian text.
A brief sample is his comment on how *quia* demonstration
can sometimes be converted to *propter quid,* where he introduces the idea of "necessary relationship" between cause and
effect. Paul states:

> It must be said that a cause is never demonstrated
> through an effect unless there be a necessary relationship
> of the effect to the cause or vice versa. Hence a cause
> and an effect sometimes have mutual necessary relation-
> ships such that, if the cause is placed the effect is placed
> also, and vice versa, as an eclipse and the interposal of
> the earth between the sun and the moon . . . Sometimes,
> however, a cause has a necessary relationship to an effect
> but not vice versa, such that when the cause is placed
> the effect is also placed, but not the other way around,
> and so in this way the effect is more common than the
> cause, as fire and the ability to heat; for if fire be posited
> the ability to heat results, but not the other way
> around.[11]

Discussing in the same context the proof of the sphericity of
the moon from its waxing and waning, Paul proposes a novel
objection associated with the theory of pin-hole images.
"There is a doubt," he writes, "that one cannot conclude a
posteriori from the curvature of the illumination to the roundness of the body, because the light projected on a wall
through a square aperture from an angular light source is circular, and nonetheless neither the aperture nor the illuminating body is circular."[12] He replies by showing that such an
argument does not universally falsify all demonstrations
based on the projection of light rays, "because then it would
follow that if an opaque object were placed close to the aperture, the light projected would appear round, and this is
false; in fact it appears with the shape of the aperture."[13]

　　Also of interest is Paul's explanation as to how there can
be demonstration of things that happen for the most part,
thereby showing that a valid demonstrative procedure exists
that is typical of natural science. He writes:

> It should be said that demonstrations are always made
> absolutely in mathematics, because when the cause is
> placed the effect universally follows. In natural things,

this is sometimes the case and sometimes not. For whenever the earth is interposed between the sun and the moon an eclipse of the moon always follows, but it does not always follow that an olive will result from an olive seed, because such a cause can be impeded; therefore, from an effect of this kind the existence of a cause can be demonstrated, but not the other way around. Thus it follows that if an olive exists its seed either exists or did exist, but it does not follow that if an olive seed exists the olive either exists or will exist, because such a cause can be impeded; although it does properly follow that if an olive seed exists, then an olive is naturally apt to come into existence. Therefore there can be science of things that are of frequent occurrence, and if this is not actual (*secundum actum*) it is nonetheless aptitudinal (*secundum aptitudinem*), which is sufficient for science; for it is not necessary that if man be a rational animal capable of laughing that he be laughing all the time, but only that he be always capable of laughter.[14]

One problem given special attention by Paul of Venice is that related to the methodological passage from effect to cause and then back again from cause to effect, because this procedure seems open to the charge of being circular proof. Exposing first Aristotle's procedure, Paul writes:

Scientific knowledge of the cause depends on a knowledge of the effect, just as scientific knowledge of the effect depends on a knowledge of the cause, since we know the cause through the effect before we know the effect through the cause. This is the principal rule in all investigation, that a scientific knowledge of natural effects demands a prior knowledge of their causes and principles.[15]

This statement, Paul recognizes, is open to the objections that knowledge of anything would seem to depend on the knowledge itself and that there would result a possible circularity (*possibilis circulatio*) in the knowing process, which is cautioned against by Aristotle in the first book of the *Posterior Analytics*. To these objections Paul replies:

In the natural [knowing] process there are three types of knowledge. The first is of the effect without any reasoning, called *quia;* the second is of the cause through knowledge of the effect, and it is likewise called *quia;* while the third is of the effect through the cause, and this is called *propter quid.* But the knowledge *propter quid* of the effect is not the same as the knowledge *quia* of the effect, and therefore the knowledge of the effect does not depend on itself but on something else. . . . There is nothing impossible about having circularity in knowledge when there are different kinds of knowing; Aristotle [reproves circularity] when the knowing is of the same kind. Hence, as has already been said, the first knowledge of the effect is *quia* knowledge, whereas the second knowledge of the same effect is *propter quid;* but *propter quid* and *quia* knowledges do not pertain to the same type of knowing.[16]

These citations occur in Paul's *Summa philosophiae naturalis,* a compendium of Aristotle's teaching in the *libri naturales,* but which incorporates into this Aristotelian context all of the significant teachings of the Oxford calculators and the Paris terminists. This work had a marked influence on later writings, and served to unify nominalist thought with the older Aristotelian traditions at Padua in a way that was almost completely lacking at Oxford and that was only imperfectly developed by the Paris terminists.

Paul's realist commitments are discernible in his commentary on the *Posterior Analytics* and also in his *Summa,* but they become more apparent in his comprehensive commentary on Aristotle's *Physics.*[17] Early in this commentary he again clarifies his teaching on the types of knowledge one may have of causes and effects, and explains how in natural science the *quia* process leads to the *propter quid.* He states:

The Commentator [Averroës] recognizes a double procedure in natural science. The first is from what is less known to nature to what is more known to nature, and is from effect to cause. The second is from what is more known to nature to what is less known to nature, and is from cause to effect . . . Natural science begins both from the causes and from what is caused, but in different senses. It begins from the causes *inclusive,* i.e., by knowing them; but from the things caused *exclusive,* i.e., by knowing by means of them . . . There is thus a twofold knowledge of every cause, the one kind by the procedure *quia* and the other by the procedure *propter quid*. The second kind depends on the first, and the first is the cause of the second; and thus the procedure *quia* is also the cause of the procedure *propter quid*.[18]

Consistent with the priorities outlined in this text, Paul turns his attention to the problem of motion and expresses a forthright commitment to the reality of movement as well as to the causes and effects involved in its production. In his commentary on Book 3, in fact, he raises the question whether local motion differs from the thing moved, the place, and the space in which it is, and answers decidedly in the affirmative.[19] His objections to the contrary are drawn from Ockham, Gregory of Rimini, and John Wyclif; to these Paul replies with a variety of arguments. The first seems indebted to the *articulos Parisienses*, and goes that God could annihilate everything in the universe but the ultimate sphere of the heavens; then, if this continued moving, it would not traverse any new space but would still continually acquire a motion distinct from itself, which must therefore be more than a mere relationship. If local motion is to be identified with the moving body, on the other hand, curvilinear motion would be rectilinear motion, and uniform motion would be difform motion, because the same identical body could be involved in each case. Paul likewise employs a number of continuity paradoxes to show that local motion cannot be an "indivisible motion," and that it cannot be a "fixed accident" (*accidens fixum*) in the object moved. Among his conclusions is the statement that local motion must be "a successive and flowing (*fluxibile*) accident" that really inheres *in* the moving object: the preposition "in" here cannot mean a relationship of predication only—it must designate a relationship of actual ontological inherence.[20] And, since local motion is a real and novel effect, it must have its own proportionate cause, and the motor causality principle is still valid for this type of motion.[21] In the teaching of Paul of Venice, therefore, one can see that the ambivalence of the Mertonians with respect to the definition of motion has been removed. Paul, in fact, is explicit that motion itself cannot be conceived merely as a ratio: "Motion is not a ratio, because a ratio is only a relative accident, whereas motion is an absolute accident." [22]

In connection with Paul of Venice mention should be made of two of his contemporaries, Jacopo da Forlì and Hugo of Siena, both of whom were concerned with methodology in the Aristotelian and Galenic traditions. Jacopo's description of resolution is interesting in this regard:

> Resolution is twofold, natural or real, and logical. Real resolution, though taken improperly in many senses, is strictly the separation and division of a thing into its component parts. Logical resolution is so called metaphorically. The metaphor is arrived at in this fashion: just as when something composite is resolved, the parts are separated from each other so that each is left by itself in its simple being, so also when a logical resolution is made, a thing at first understood confusedly is understood distinctly, so that the parts and causes touching its essence are distinctly grasped. Thus if when you have a fever you first grasp the concept of fever, you understand the fever in general and confusedly. You then resolve the fever into its causes, since any fever comes either from the heating of the humor or of the spirits or of the members; and again the heating of the humor is either of the blood or the phlegm, etc.; until you arrive at the specific and distinct cause and knowledge of that fever. [23]

Jacopo thus illustrates his methodology with examples drawn from the practice of medicine. Himself quite expert in calculatory techniques, it is noteworthy that Jacopo wrote a treatise on the intension and remission of forms that justifies a quantitative and mathematical approach to this subject, as opposed to the merely qualitative logic of intension and remission that had been developed by Walter Burley. [24]

Hugo of Siena insists, in a fashion somewhat similar to Jacopo, that any *doctrina* must involve a setting forth of what is demonstrable, a *manifestatio demonstrabilis*, and that as practiced in physics and medicine it must involve two processes, the first of which begins with the effects and seeks their cause and the second explains those effects through the newly discovered cause. Hugo writes:

In the discovery of the middle term or cause we proceed from effect to cause . . . Such a way of acquiring knowledge we call resolutive, because in that discovery we proceed from an effect, which is commonly more composite, to a cause which is simpler; and because by this discovery of the cause we certify the effect through the cause, we say that demonstration *propter quid* and "of the cause" is the foundation of resolutive knowledge . . . But I myself see in the discovery of a science of effect through cause a double form of procedure, and likewise in the discovery of a science of cause through effect. The one procedure is the discovery of the miʲdle term or cause, the other is the setting forth of its consequences or effects. And the process of discovery in the case of demonstrations through causes is resolutive, while that of setting forth the consequences is compositive. In demonstration through effects it is just the other way around.[25]

What is important in this text, as Randall has noted, is again the linking of the process of discovery with that of proof.[26] For Hugo, as for others of the Paduan School, both discovery and proof are essential components of a strictly scientific procedure.

b. Gaetano da Thiene

The culmination of the work done by Paul of Venice during the fifteenth century can be seen in the writings of his most important disciple, Gaetano da Thiene.[27] Gaetano was not only a consummate Aristotelian in the scholastic and Averroist traditions but he also understood and had great sympathy for the kinematics of the Mertonians and the dynamics of the Paris terminists. Thus he commented on the *libri naturales* of Aristotle, but he prepared also a discerning commentary on the *Regule* of William Heytesbury.[28] It is to the latter that we should turn our attention now for the indication it provides of a change of mentality from fourteenth-century Oxford to fifteenth-century Padua.

The Mertonians, as we have indicated, viewed local motion as essentially a mathematical ratio, and thus prepared the way for fairly sophisticated treatments of the relationships that obtain between velocity, time of travel, and distance traversed. They conceived of a great variety of motions as taking place in imaginary space; when they came to discuss motive forces and resistive media they did so in a purely mathematical way, *secundum imaginationem*, and without any special reference to the real world. This attitude was a natural outgrowth of the Ockhamist view of motion, which did not see it as an independent entity requiring its own causes and effects, but rather as a concept to be treated logically in terms of the various sophisms that arose when one did not use language properly in its discussion. As opposed to this basically nominalist orientation, the realist position, such as we have seen beginning to develop at Paris, was directed primarily toward an understanding of the real world, the world of nature. Thinkers such as Paul of Venice would indulge in highly imaginative treatments of motion, and they were not adverse to the use of mathematics in so doing, but they were quite opposed to any simple equating of motion with a quantitative ratio. They used the complex terminology of the nominalists relating to the latitude of forms, for example, but their concern was not with quantitative definitions alone. Since they regarded motion as a real entity requiring its own causes and producing its own effects, they could not help but inquire into the dynamical factors associated with its analysis. And rather than being concerned with merely imaginative examples, they sought cases in the order of nature that would exemplify the kinematical definitions of the Mertonians, and then sought to connect these with the dynamical ideas proposed by the Paris terminists.

An instance of this trend can be seen from reading the following small part of Heytesbury's *Regule* and then contrasting this with the comment provided by Gaetano. Heytesbury points out at the beginning of his treatment of local motion the distinction between uniform and non-uniform motion. He then explains how one goes about measuring a uniform velocity:

> In uniform motion . . . the velocity of a magnitude as a whole is in all cases measured by the linear path traversed by the point which is in most rapid motion, if there is such a point. And according as the position of this point is changed uniformly or not uniformly, the complete motion of the whole body is said to be uniform

or difform. Thus given a magnitude whose most rapidly moving point is moved uniformly, then, however much the remaining points may be moving non-uniformly, that magnitude as a whole is said to be in uniform movement . . .[29]

The language, as is easily recognized, is that of kinematics. Heytesbury is writing of moving points, but these he conceives very abstractly and one is hard put to see how his ideas apply in any way to the order of nature.

Commenting on this section, however, Gaetano's exposition takes a realistic and practical turn. To exemplify Heytesbury's reasoning he proposes the case of a rotating wheel that expands and contracts during its rotation. He talks also of a cutting edge placed against a wheel that continually strips off its outermost surface. Another of his examples is a wheel whose inner parts are expanding while its outer surface is being cut off. Gaetano speaks too of a disk made of ice rotating in a hot oven; here the outermost surface continually disappears and the velocity at the circumference becomes slower and slower, whereas the inner parts expand under the influence of heat and their linear velocity increases. Yet another of his examples is a wheel that rotates and has material gradually added to its circumference, as clay is added by a potter to the piece he is working. Here the velocity of rotation would be uniform but the linear velocity of a point on the circumference would increase, unless the entire wheel could be made to contract in the process, in which case the linear velocity of the outermost point might remain constant.[30]

These examples, it should be noted, are Gaetano's and not Heytesbury's. Heytesbury's kinematic doctrine is, of course, important, for without it Gaetano would have had no reason to seek its exemplification. But the examples furnished by Gaetano are important too, for they show that Gaetano was convinced that Heytesbury's doctrine could be applied to the real world, and in fact was thinking of cases that were realizable in materials close at hand after the fashion of the experimenter. Gaetano did not perform experiments or measurements, so far as we know, but he took another step closer to their realization. And he, like Paul of Venice, was a realist,

in the sense of the moderate realism of scholasticism, but
nonetheless unwilling to accept fully the nominalist philosophy of nature.[31] He also subscribed to the resolutive and
compositive process in natural philosophy,[32] which we have
seen to be quite developed in the Paduan School, and he was
disposed on this account to seek the causes of natural phenomena, not the least of which were those associated with
local motion.

The mention of experimentation suggests at least a passing reference to the Milanese physician, Giovanni Marliani,
who studied and taught at the University of Pavia and who
composed a *Quaestio de proportione motuum in velocitate*, in
which he mentions having lectured on the treatises on ratios
written by Bradwardine and Albert of Saxony.[33] In proposing
various arguments relating to Bradwardine's function, most of
which he recognized as erroneous, Marliani invokes an experimental type of reasoning based on the use of pendulums and
the rolling of spherical weights down inclined planes. Randall and Clagett disagree on the precise significance of these
experiments; [34] regardless of such differences of interpretation, however, the mere mention of these tests reveals the
mentality toward natural science that was then developing in
northern Italy and that ultimately would result in Galileo's
achievement. Again, whatever may have been Marliani's personal competence, he does cite an impressive list of sources,
including Bradwardine, John of Dumbleton, Heytesbury, Buridan, Albert of Saxony, Nicole Oresme, and Paul of Venice.[35]
Thus the growing body of knowledge relating to kinematics
and dynamics that had been acquired at Oxford, Paris, and
Padua was already diffused throughout northern Italy, await-
ing only a development along empiricist and experimental
lines to produce the "new science" that would emerge in the
seventeenth century.

2. *Revival at Paris*

Before this could be accomplished, a further theoretical development would prove helpful, and this came from outside
Italy although it was at least partially inspired by the work of
the Paduan calculators. The development concerns the con-

cept of motion as this came to be understood in the sixteenth century, when new attempts were made at synthesis and at a reconciliation of the divergent speculative views of the nominalists and the realists. The center at which this reconciliation took place was the University of Paris, and the man most responsible for it, in Hubert Élie's analysis, was the Scottish scholar, philosopher, and theologian, John Major of Haddington, known in Latin as Joannes Maior, or, more generally, by the French version of his name, Jean Mair.[36] By accident or design, Mair numbered among his students Scots, Belgians, French, Germans, Spaniards, and Portuguese; the only important nation not represented was Italy, and this because her universities already constituted another pole of attraction for scholars from all over Europe. But if Italians were not themselves present at Paris, their works were, for practically all of the writings of the Paduans we have mentioned were known and discussed by Mair and his disciples. The resulting literary output at sixteenth-century Paris was quite extensive,[37] so we shall limit our attention here to commentaries and "questionaries" on the *Physics* of Aristotle for the indications these provide of a changing concept of motion.

a. Jean Mair

Jean Mair is usually identified as a nominalist, but in his treatment of motion he does not subscribe wholeheartedly to the Ockhamist position.[38] In fact, his way of approaching the reality of motion is very similar to that of Albert of Saxony, for Mair also has two questions on Book 3 of the *Physics* dealing with this topic.[39] Both raise exactly the same query, "Whether local motion is a successive entity that is distinct from anything permanent?" Question 2 notes that "there are sides" on this, and the first is that "of the realists" who hold for the affirmative. Mair thereupon explains the realist solution, and then raises nine objections against it, some of which are theological; each objection he considers in turn, and explains how it can be answered "by this school (*secundum hanc viam*)," thus concluding the question. Question 3 then follows immediately, and treats the same problem, only now from the negative side, for those who regard the difficulties raised in the objections as not being adequately solved. Mair notes that there are various schools, too, on the negative, meaning by this the opinions of Gregory of Rimini and of William of Ockham, which he explains, and then raises thirteen objections to the contrary, including the "divine cases proposed by Albert of Saxony." He concludes by answering all thirteen objections, and then goes immediately to the next question, without any comment about, or justification of, his seemingly eclectic procedure.[40] That he is sincere in seeking some truth on both sides, however, and in regarding the difference as mainly terminological, seems indisputable. And Mair is consistent in his later treatment of motion, for he considers in subsequent books of the *Physics* all the topics relating to the kinematics of motion that were customarily discussed by the nominalists, as well as dynamical problems, such as "how the velocity of local motion is ascertained from its cause," that exhibit realist concerns.[41]

b. Dullaert and Coronel

Mair's Flemish disciple, Jean Dullaert of Ghent, an Augustinian who himself edited the works of Paul of Venice, shows the same dualistic tendency as his teacher, although he goes into the problem in more detail, devoting some twenty pages to it.[42] He poses the question, "Whether motion is something successive distinct from any permanent thing," and replies:

> On this there are various opinions, and first, beginning with local motion, there are many opinions as to what it is. Some "reifiers" say that local motion is one accident really inhering in the movable body. And these are further divided. Some say it is a "respective" accident — Burley follows this view; others say that it is an "absolute" accident, and Paul of Venice takes this position. Still others, like the nominalists, deny that local motion is such a successive accident, and these too are further divided. Some, like Gregory of Rimini, hold that local motion is the space itself over which the movable body moves; others say that local motion is only the movable object.[43]

With this statement, Dullaert first defends the realist positions, citing Buridan and Paul of Venice, and, in one place, accusing "almost all the nominalists" of inconsistency and stupid argument.[44] Then, without explanation or apology, he briefly exposes and defends Gregory of Rimini's position, noting at the end that "few nominalists" follow this, and so he goes on to present a third opinion. This last exposition, likewise fairly brief, cites Albert of Saxony and George of Brussels, and concludes with the summary statement:

> Among these opinions, the first is more subtle and consonant with the sayings of the Philosopher [Aristotle]; the second is less popular; and the third is regarded as true and is more common among the moderns.[45]

Thereupon follows an extensive analysis of ratios, required for studying the "velocity of local motion," and then a full exposition of the teachings of Swineshead, Heytesbury, and Oresme on the intensities of forms and the velocities of alteration and augmentation.[46]

The ambivalence of Mair and Dullaert is also discernible in the *Physice perscrutationes* of Luis Nuñez Coronel, one of Mair's early Spanish disciples.[47] Coronel is also of interest, as Duhem has pointed out, for his methodological use of the expression "saving the phenomena," and this in a context where most Aristotelians would have resorted to the procedures of resolution and composition.[48] At the outset of his exposition of Book 1 of the *Physics*, Coronel attempts to show that every extended substance is composed of a substantial form and a [primary] matter that is in no way producible by natural activity.[49] In listing objections to his thesis he argues that many natural phenomena "can be saved" without the necessity of positing such a matter. Answering this objection, Coronel notes as an experimental fact (*cognoscimus per experientiam*) that we cannot start a fire without destroying something combustible, and concludes from this that matter is actually indispensable because the phenomenon of burning could not be accounted for if fire were pure form. He then explains his method of reasoning, insisting that in physics one must proceed "from those things that are known by experiment (*experimento*)," and goes on:

> As Albert the Great maintained, in the discipline of physics arguments drawn from experience fill the role of

principles. The arguments of astronomers, drawn from
the diversity of the celestial motions and the distances
between the heavenly bodies and the planets, led to the
proposal of epicycles, eccentrics, and deferents as conclusions. Similarly, matter must be posited, as required by
natural reason. For if it is not, the fact that making a fire
requires the supply of something combustible cannot be
saved (*non potest salvari*), just as the celestial appearances cannot be saved unless one posits epicycles, etc.
The hypothesis of epicycles, eccentrics, and deferents of-
fended the Commentator, Averroës, but he provided no
alternative method of saving what is saved by these assumptions. Moreover, the same might be said of matter
and the other causes which he himself admitted to save
the things that occur naturally.[50]

The overly realistic view of eccentrics and epicycles implied
in this citation does not seem to be completely consistent
with the remainder of Coronel's philosophy, but it provides
an instance of how hypothetical reasoning was used by at
least one sixteenth-century author at Paris to justify the basic
principles of his natural philosophy.

When treating the question of the reality of motion, Coronel lists the three by then classical positions, namely, those
of the realists, the less popular nominalists, and the more
popular nominalists, and declares his intention first to defend
all three and then to give his judgment as to which is "more
probable." [51] In his exposition of the three views he cites
many authorities: Scotus, Buridan, Paul of Venice, Jacopo da
Forlì, Burley, Peter d'Ailly, Gregory of Rimini, and George of
Brussels. Finally, he appends two sections to his exposition,
the first devoted to how local motion produces heat (using arguments from Thomas Aquinas), and the second to a lengthy
treatment of his personal views on impetus. Then he concludes:

Having exposed and defended the various views concerning local motion, with some omissions, there remains the
task of selecting the "more probable" view. But this I
leave to the judgment of others. The first position is older
and [more] subtle; the second is extraneous and uncom-
mon; the third is easier and better appearing. The fourth

(which we did not wish to enumerate at the outset) is intelligible and not completely improbable — it satisfies very well the three arguments we raised against the third position, and is no less able to explain the heat resulting from motion, the "aptitude" left after motion, and the immovable impetus produced. And this suffices for the first part of this third book.[52]

The so-called "fourth position" is not completely clear, although it seems to propose a teaching intermediate between the realist and the more common nominalist view. It is, moreover, the first explicit indication we have of a new view of the entitative status of local motion to emerge in the sixteenth century. Yet, despite this innovation, Coronel does not exhibit great interest in the problems associated with local motion. He discusses at length the intensification of qualities and the latitude of forms, in a manner reminiscent of Oresme, but when he comes to treat the velocity of motions, he is extremely brief: "We proceed very briefly and succinctly in this disquisition, because I do not think it worthwhile to dwell on such matters." [53]

c. Celaya and Soto

The suspicion that some kind of rapprochement was emerging from the ambivalent treatments of Jean Mair and his disciples is possibly confirmed by the title of another Parisian master, also a Spaniard, but not a direct disciple of Mair. This is Juan de Celaya, who wrote his *Expositio in octo libros physicorum Aristotelis* at Paris in 1517, appending the subtitle "With Questions According to the Three Schools of St. Thomas, the Realists, and the Nominalists." [54] What is significant about Celaya is not only his mention, in the subtitle, of a Thomistic position as a *tertia via* different from those of the realists and the nominalists, but also his numbering among his students a Spanish layman, Francisco de Soto, who was later to become a Dominican friar and take the name of Domingo. Like Coronel, however, in the final analysis Celaya is eclectic; he rests content with enumerating the different positions, without taking sides, and supplies a compendious treatment of all matters that would interest a nominalist, a realist, or anyone inclined to see elements of truth in either position.[55]

The culmination of this sixteenth-century development at Paris relating to motion comes in a work of Domingo de Soto, *Super octo libros physicorum Aristotelis quaestiones,* composed at the University of Salamanca *c.* 1545, but undoubtedly based on lectures by Soto at the University of Alcalá in the early 1520's, shortly after he had returned from Paris.[56] In his second question on Book 3, Soto inquires "Whether motion is something distinct from the thing moved and the form or terminus [attained]?" and exposes, in the Parisian manner, both the realist and the nominalist replies with their better known variations.[57] His own answer is that both answers contain elements of the truth, and that the difference between the realist and the nominalist positions is mainly one of terminology. If one wishes to restrict the notion of real distinction to the differentiation of substances that are numerically distinct, then local motion is not really different from the object moved or from the location reached. Yet, even though all of these exist "identically" in the same subject, they are not to be formally identified, since each has a different *ratio* or definition. At the least they are different in the mind's way of considering them, even though they exist in one and the same body. Soto is even willing, so as to avoid further dispute, to call the distinction that Aquinas and the older Aristotelians referred to as a real modal distinction, merely a "distinction of reason." This, he thinks, is closer to the "connotations" that are spoken of by the nominalists, and a "distinction of reason," when properly understood, is sufficient to save the different ways of speaking about local motion, the object moved, and the space traversed. But Soto would avoid both the realist and the nominalist extremes: each "sins through excess," in his estimation. He does not believe one should multiply entities, but neither should one dispense completely with the Aristotelian categories — without them meaningful discourse becomes impossible. And he is explicit that motion itself, while only rationally distinct from the object moved, is not on this account to be regarded as a mere *ens rationis.*[58] Like a quality, it does require a cause, and it does produce distinctive effects, so that the motor causality principle still applies to local motion.[59]

The outcome of Soto's analysis is that it enables him to take a consistent theoretical position with regard to motion and its properties. This position is not eclectic, as was that of Soto's immediate predecessors, although it recognizes elements of truth in both the nominalist and the realist extremes. More important, it provides a workable basis for a consistent treatment of motion in both its kinematic and dynamic aspects. Soto, having eliminated the logical problems of the *sophismata*, can still treat the quantitative aspects of local motion, and he does so in his "digression on ratios" and in his analysis of how the velocity of motion is to be ascertained "from its effects." [60] This is the standard kinematical treatise of the Mertonians, only with this difference, that Soto presents it, not as an abstract and imaginative mathematical exercise, but rather as an analysis that applies to motions in the physical universe. Moreover, this same concept of motion gives him a consistent reason for also ascertaining the velocity of motion "from its cause," [61] and thus for taking up also the dynamical questions that were to be hotly debated in northern Italy by Galileo's predecessors and contemporaries, and by Galileo himself in both his Pisan and Paduan periods.

Soto is best known to historians of science for having, as Duhem pointed out, applied the mean-speed theorem to the case of free fall, and thus for having adumbrated the so-called "law of falling bodies" more than half a century before Galileo.[62] The context in which this application is made is in a question on Book 7 of the *Physics* where Soto is discussing the various types of uniform and difform motions.[63] For some curious reason, when exemplifying these motions most writers used a system of classification that may be traced back to Albert of Saxony.[64] This included, among others, motions that are uniform with respect to time and difform with respect to the parts of the moving object, and motions that are difform with respect to time and uniform with respect to the parts of the moving object. The first was by Soto's day commonly exemplified by a wheel or by a heavenly sphere, which rotates uniformly with respect to time but whose parts move with greater velocity as they are located farther from the center or the pole and closer to the outermost periphery. The second was similarly exemplified by the falling body, whose velocity of fall increases with time, but all of whose parts move with the same velocity at any instant. With very few exceptions, the authors before Soto who attempted to illustrate uniform or difform motions, and these would include Nicole Oresme, Gaetano da Thiene, and John Dullaert of Ghent, did so with examples that employed this two-variable schema. They always spoke of variations that take place both with respect to time and with respect to the parts of the object moved, and spoke of either being uniform or difform in all the possible combinations.[65]

Soto's advance here, it would seem, was one of simplification. He thought of discussing motion that is uniform merely with respect to time, or uniform merely with respect to the parts of the object moved, and gave simple illustrations of these. He exemplified also motions that are difform with respect to time alone, and then went further to seek examples from nature illustrating how some motions are uniformly difform with respect to time, whereas others are difformly difform in the same respect. In other words, Soto substituted a one-variable schema for a two-variable schema, and restricted himself to one variable at a time when furnishing realistic, as opposed to imaginary, examples. Thus, when seeking examples to illustrate local motions that are uniformly difform with respect to time, Soto states that these are "properly found in objects that move naturally and in projectiles."[66] He goes on:

> For when a heavy object falls through a homogeneous medium from a height, it moves with greater velocity at the end than at the beginning. The velocity of projectiles, on the other hand, is less at the end than at the beginning. And what is more, the [motion of the] first increases uniformly difformly, whereas the [motion of the] second decreases uniformly difformly.[67]

Later on in the text Soto removes any possible ambiguity as to whether he means that the motion is uniformly difform with respect to the time of fall or with respect to the distance of fall by proposing the difficulty "whether the velocity of an object that is moved uniformly difformly is to be judged from its maximum speed, as when a heavy object falls in one hour with a velocity increase from zero to eight, should it be said to move with a velocity of eight?" [68] His answer to this is clearly in terms of the Mertonian mean-speed theorem, for he decides in favor of the average velocity (*gradus medius*) as opposed to the maximum. He justifies this with the illustration: "For example, if the moving object *A* keeps increasing its velocity from zero to eight, it covers just as much space as [another object] *B* moving with a uniform velocity of four in the same [period of] time." [69] Thus there can be no doubt about his understanding of *uniformiter difformis* and how this is to be applied to the space traversed by a freely falling object.

Did Soto ever measure the distance covered by a falling body to see if his exemplifications were correct? Certainly he did not. There are indications in his writings that he performed what later thinkers would call "thought experiments" (*Gedankenexperimenten*), particularly relating to the vacuum, but he seems not to have done any measuring or experimenting himself.[70] What is significant about his contribution is that he laid the groundwork, that he prepared the ideas, that he simplified the examples, so that someone else might see that here was a case that is experimentally tractable, and finally put a mathematical law of motion to empirical test.[71]

3. *Sixteenth-Century Padua*

The final chapter in this development, of course, was written at the turn of the century in northern Italy, at Padua, fittingly perhaps, in view of the work done there earlier by Paul of Venice and Gaetano da Thiene. Here Averroist and scholastic Aristotelianism continued to be a potent force in the universities, although there was also increasing interest in Platonism, and the application of mathematics to physical problems along lines suggested by Archimedes was beginning to attract attention. The culmination of the Paduan tradition we have been discussing, however, came about in the writings of Agostino Nifo and Jacopo Zabarella, and to these we will first give our attention.

a. Agostino Nifo

Agostino Nifo is perhaps best known for his treatise on the immortality of the human soul directed against Pietro

Pomponazzi, wherein he reveals an interest in Plato as well as in Aristotle.[72] His major concern was with the latter, predictably, for he edited the works of Averroës and composed numerous commentaries on Aristotle that earned for him a reputation as one of the best peripatetics of his time. Of special interest are his commentaries on the *Posterior Analytics* and the *Physics,* in both of which he sheds light on the tradition whose development we have been examining.

The fifteenth-century emphasis on the double procedure of resolution and composition came to be known at Padua by the term *regressus,* and was commonly distinguished from circular reasoning (*circulatio*) on the basis of the explanation we have seen in Paul of Venice. In his commentary on Book 1 of the *Physics,* Nifo explains Averroës's teaching on the three kinds of demonstration and on the twofold process involved in the study of natural things, "one from the effect to the discovery of the cause, the other from the cause discovered to the effect," and then turns to the question whether such a twofold process really involves circular proof (*circulus in demonstrationibus*).[73] In answer, he analyzes the commentaries of Themistius, Philoponus, and Averroës, showing in the latter case that the *regressus* is not circular. He then explains:

> Recent writers (*recentiores*) maintain that there are four kinds of knowledge. The first kind is of the effect through sense or observation; the second is the discovery of the cause through the effect, which is called demonstration "of sign"; the third is knowledge of the same cause through the work (*negotiatio*) of the intellect, from which with the first there comes such an increased knowledge of the cause that it is rendered fit to serve as the middle term of a strict demonstration; the fourth is a knowledge of that same effect *propter quid* through that cause known so certainly as to be a middle term.[74]

Since the second knowledge of the effect is quite different from its initial observation, there is no circular argument but rather a true regress. Of key significance in Nifo's explanation is what he describes as the *negotiatio* of the intellect, and this he feels obliged to explain:

> This *negotiatio* is composition and division. For when the cause itself has been discovered, the intellect com-

poses and divides until it knows the cause precisely as a middle term (*sub ratione medii*). For, though the cause and the middle term are the same thing, they differ in understanding (*ratio*). For something is called a cause in-asmuch as an effect proceeds from it, whether it be bet-ter known than the effect or not, whereas it is a middle term inasmuch as it is a definition. The procedure of dis-covering the cause is thus from effect to cause, and *nego-tiatio* is a grasping of the cause as a middle term and definition. But since a definition is discovered only by composition and division, it is through them that the cause is discovered under the formality of a middle term, from which one may then proceed to the effect.[75]

The foregoing was written by Nifo in 1506, and his ex-planation is similar to that offered by Paul of Venice and other scholastics, particularly in its attributing to the human intellect the power to discern causes with the degree of certi-tude necessary to generate *propter quid* demonstration. Com-posing a *Recognitio* on this same text several decades later, however, Nifo has doubts about this power of the intellect to generate certitude, and revises his thinking along hypotheti-cal lines that are more in accord with the methods of modern science. First he surveys what has been recently written on the "regress," to indicate that what was previously thought to imply four types of knowledge is now considered as involving only three.[76] Having done this, Nifo apparently returns to his former explanation where he has singled out the work (*nego-tiatio*) of the intellect for special mention, and proceeds to re-vise his opinion:

From this it is clear that there is no need of any *negotia-tio* to render greater our knowledge of the cause, as we formerly held; for the mere knowledge that it is the cause (*quia est*) is the *propter quid* of the effect. Yet when I more diligently consider the words of Aristotle, and the commentators Alexander and Themistius, of Philoponus and Simplicius, it seems to me that in the regress made in physical demonstrations the first process, by which the discovery of the cause is put into syllogistic form, is a mere hypothetical (*conjecturalis*) syllogism, since through it the discovery of the cause is syllogized in a merely conjectural fashion. But the second process, by

which is syllogized the *propter quid* of the effect through
the discovered cause, is demonstration *propter quid* —
not that it makes us know *simpliciter,* but conditionally
(*ex conditione*), provided that that actually *is* the cause,
or provided that the propositions are true that represent
it to be the cause, and that nothing else can be the
cause.[77]

Of special interest in this passage is Nifo's use of the expression *syllogismus conjecturalis,* or hypothetical syllogism, for
indicating what in the Aristotelian tradition had come to be
known as a posteriori demonstration. In his commentary on
the *Posterior Analytics* Nifo explains at length his understanding of the difference between *demonstratio simpliciter*
and *demonstratio conjecturalis,* and gives as an example of
hypothetical demonstration the proof of the moon's sphericity
from its waxing and waning.[78] He maintains that this is a true
propter quid demonstration, but not absolutely speaking
(*simpliciter*), "because the discovery of the cause is not absolutely obvious to us, but it is conjecturally (*conjecturabiliter*),
as Philoponus says. For this reason the regress is a demonstration *ex hypothesi,* 'if that *is* the cause,' because it is not
absolutely obvious that it is actually so." [79] The same line of
thought is contained in the *Recognitio* added to the *Physics*
commentary, for there, having introduced his distinction between demonstration *simpliciter* and *conjecturalis,* Nifo raises
the problem whether in this way he has not destroyed the
very possibility of scientific knowledge of the world of nature:

But you object that in that case the science of nature is
not a science at all. We must say that the science of nature is not a science *simpliciter,* like mathematics. Yet it
is a science *propter quid,* because the discovery of the
cause, obtained through a conjectural syllogism, supplies
the *propter quid* of the effect . . . That something is a
cause can never be so certain as that something is an effect, for the effect's existence is known to the senses. That
it is the cause remains conjectural, whether or not such a
conjecture is more known than the effect itself in the
order of knowledge *propter quid.* For if the discovery of
the cause is assumed, the *propter quid* of the effect is always known. Hence in the *Meteors* Aristotle grants that

he is not setting forth the true causes of natural effects, but only insofar as was conjecturally possible for him.[80]

Much the same teaching is contained in Nifo's commentary on the *Posterior Analytics,* where he makes the distinction between *demonstratio simpliciter* and *demonstratio conjecturalis,* requiring for the former that the cause be more known to nature and also to us, and noting that "all mathematical demonstrations and some physical demonstrations are of this kind."[81] In the case of the conjectural demonstration the cause would be less known to us than the effect, and it is really demonstration "from the hypothesis, 'if that *is* the cause,' because it is not simply apparent (*non simpliciter patet*) that it is so."[82] In another place, however, Nifo states that some conjectures are necessary (*necessariae*), and he instances that involved in explaining the phases of the moon, whereas others are "merely rhetorical" (*solum rhetoricae*), and apparently revisable on this account.[83] Nifo, to my knowledge, does not give an example of a *demonstratio simpliciter* in physical science, but seemingly this would be through a cause or principle that would not be discovered merely by a reasoning process but would be evident to the senses, the same as are the effects with which most physical demonstrations begin.

The significance of Nifo's development of the Aristotelian theory of demonstration, as has been pointed out, is that it already provides, "at the beginning of the sixteenth century . . . a clear formulation of the structure of a science of hypothesis and demonstration, with the dependence of its first principles upon empirical investigation plainly set forth."[84] By elaborating his teaching in this distinctive way, Nifo is able to bracket, as it were, the question of absolute certitude and its attainability through physical demonstration. Earlier scholastics, such as Aquinas with his definition of science as *certa cognitio per causas,*[85] had stressed the absolute and invariant character of demonstrative knowledge, although, as we have seen, Aquinas himself gave sufficient indication that demonstrations *ex suppositione* were more the rule than the exception in natural science. In connection with Aquinas, it is noteworthy that Nifo states in his *Physics* commentary that the physical scientist "sometimes uses all four causes, sometimes several, and sometimes only one," [86] and that he is explicit on the importance of demonstrations through final causality in natural science, understanding this much in the way explained by Aquinas.[87] He is also markedly realist in his interpretations of the *Physics,* as were Renaissance Aristotelians generally, and thus did not subscribe to nominalist doctrines, although he was aware of their existence. Motion for him was real, and it could be viewed either as a *fluxus formae* or as a *forma fluens.* "Thus I say that *fluxus* and *forma fluens* are one thing, although they differ in understanding (*ratio*), just as do *fluxus* and *forma remissa.* For when a movable object is not at rest under any remiss form, there is a *fluxus* or *forma fluens;* whereas when it is lacking the ultimate perfection of its species, it is spoken of as a *forma remissa.*" [88]

b. Jacopo Zabarella

Apart from Nifo other Renaissance Aristotelians, such as Alessandro Achillini, Marc Antonio Zimara, and Bernardinus Tomitanus, developed the logic of proof during the sixteenth century,[89] but it remained for Jacopo Zabarella to formulate the classical version and to make current the methodological terminology that was to be used by the great scientific discoverers who emerged from the School of Padua, Galileo included. A professor at Padua from 1564 to 1589, Zabarella wrote numerous works on logic, including an extensive commentary on the *Posterior Analytics,* and some treatises on natural philosophy. He was one of the most respected of the Aristotelian commentators in his day, and later exerted considerable transmontane influence, particularly in Germany with J. Jungius and his celebrated student G. W. Leibniz.[90]

For Zabarella logic is practically identified with method, and science itself is "nothing more than logical method put to use." [91] Again, "the definition of method does not differ from the definition of the syllogism." [92] A scientific syllogism, or demonstration, involves the notion of cause, and also that of "essential or necessary connection (*connexus essentialis ac necessarius*)." [93] Thus Zabarella writes:

> For all scientific progress from the known to the unknown is either from cause to effect or from effect to

cause. The former is the demonstrative method, the latter the resolutive; there is no other procedure which generates a certain knowledge of things. For if we progress from something to something else, of which neither is the cause of the other, there cannot be between them any essential and necessary connection; hence no certain knowledge can follow from that progress. It is thus clear that there can be no scientific method except the demonstrative and the resolutive.[94]

The demonstrative method, effectively that of composition, is most appropriate in mathematics, while the resolutive method is characteristic of the natural sciences, where one must start from effects because the causes are unknown. Zabarella continues:

Since because of the weakness of our mind and powers the principles from which demonstration is to be made are unknown to us, and since we cannot set out from the unknown, we are of necessity forced to resort to a kind of secondary procedure, which is the resolutive method that leads to the discovery of principles, so that once they are found we can demonstrate the natural effects from them. Hence the resolutive method is a subordinate procedure, and the servant of the demonstrative.[95]

He then goes on to explain that the difference between the two methods is explicable in terms of their ends:

The end of the demonstrative method is perfect science, which is knowledge of things through their causes; but the end of the resolutive method is discovery (*inventio*) rather than science, since by resolution we seek causes from their effects that we may afterwards know the effects from their causes, not that we may rest in a knowledge of the causes themselves.[96]

Having thus set the stage for his discussion of resolutive method, Zabarella, pursuing a line of thought that was only implicit in Nifo, points out that there are actually two methods of resolution:

The one is demonstration from effects, which in the performance of its function is exceedingly efficacious; and we employ it for the discovery of those things that are very obscure and hidden. The other is induction, which is a much weaker form of resolution, and is employed for the discovery of only those things which are hardly unknown but yet need to be made a little clearer.[97]

At this point Zabarella explains how the two methods of discovery and induction differ from each other, and then he goes on to note the types of principles that are attainable by resolution, and particularly by the method of induction:

By these differences then induction is distinguished from demonstration through effects. For each is a resolutive method of going from consequent things to principles. But two kinds of principle are presented to us. The one kind is known *naturaliter* and hence needs no logical instrument except induction, by which alone such principles can be made known. For all our knowledge takes its origin from sense, nor can we know anything with our minds unless we have known it first by sense. Hence all principles of this kind are made known to us by induction, and are therefore not said to be demonstrated or proved; for those things only are said to be proved, strictly speaking, which are demonstrated through something else. But induction does not prove a thing through something else; in a certain sense it reveals that thing through itself. For the universal is not distinguished from the particular in the thing itself but only by reason. And since the thing is better known as a particular than as a universal, because it is said to be sensible as particular and not as universal, induction is thus a process from the same thing to the same thing — from the same thing in that aspect in which it is more obvious, to the knowing of that same thing in that aspect in which it is more obscure and hidden. Therefore not only are the principles of things known by induction, but also the principles of science or of knowing itself, which are said to be indemonstrable.[98]

This careful analysis of the inductive process actually outlines an analytic method of discovery whereby ordinary experience is brought to the level of the scientific. In this process, as Zabarella indicates in his treatise *De regressu*, it is not necessary that every fact or particular be recorded, since the general principle can be gotten inductively by a careful examination of selected instances or illustrations. This presupposes, of course, that there is a basic intelligibility in the subject matter under consideration.

> Demonstrative induction can be carried on in a necessary subject matter, and in things that have an essential connection with each other. Hence it does not take all the particulars into account, since after certain of them have been examined our mind straightaway notices the essential connection, and then disregarding the remaining particulars proceeds at once to bring together the universal. For it knows that it is necessary that the same relations should be embodied in the rest.[99]

This citation shows that Zabarella has confidence in the ability of man's intellect to grasp an explanatory principle from a careful study of its consequents. One might wonder, at this point, whether he was aware of Nifo's early attempt to discuss the work (*negotiatio*) of the intellect and then his later relinquishing of this notion. Apparently Zabarella was aware of this problem, and himself amplifies and corrects Nifo's teaching without mentioning him by name. He does this also in the *De regressu,* after he has explained the four stages involved in the regressive process. He writes:

> When the first stage of the procedure has been completed, which is from effect to cause, before we return from the latter to the effect, there must intervene a third intermediate work (*labor*) by which we may be led to a distinct knowledge of that cause which so far has been known only confusedly. Some thinkers [e.g., Nifo] knowing this to be necessary have called it a *negotiatio* of the intellect. We can call it a mental examination of the cause itself, or a mental consideration. For after we have hit upon that cause, we begin to consider it, so that we may also understand what it is (*quid sit*).[100]

Zabarella then goes on to explain the nature of this mental consideration:

> But what this mental consideration may be, and how it is accomplished, I have seen explained by no one. For though some say that this intermediate *negotiatio* of the intellect does play a part, still they have not shown how it leads us to distinct knowledge of a cause, and what is

the precise force of this *negotiatio*. Thus we think it
would be well worth the effort for us to say something
about this matter. There are, I judge, two things that
help us to know a cause distinctly. One is the knowledge
that it is (*quod est*), which prepares us to discover what
it is (*quid sit*). For when we have some foreknowledge
about anything, we are able to search out and discover
something else in it; when we have no foreknowledge at
all, we shall never discover anything . . . Hence when
we find that a cause is suggested, we are in a position to
seek out and discover what it is. The other help, without
which this first would not suffice, is the comparison of
the cause discovered with the effect through which it
was discovered, not indeed with the full knowledge that
this is the cause and that the effect, but simply comparing this thing with that. Thus it comes about that we are
led gradually to the knowledge of the conditions of that
thing; and when one of the conditions has been discovered we are helped to the discovery of another, until we
finally know this to be the cause of that effect.[101]

From this rather long text it can be seen that Zabarella disa-
grees to some extent with Nifo in the sense that he does not
emphasize the hypothetical or conjectural aspect of demonstration in the physical sciences. Although he is seemingly
aware that a considerable amount of dialectical inquiry may
be necessary, Zabarella has confidence in the ability of the
human mind to discover, upon careful consideration and
analysis, the causes of natural phenomena. And while Nifo,
with his more cautious attitude, is closer to the thought of
twentieth-century scientists on the subject of certain proof,
Zabarella was closer to the seventeenth-century founders of
modern science, who, as we shall see, shared precisely his
conviction in their search for the "true causes" behind the appearances.[102]

Zabarella wrote little about the use of mathematical reasoning in natural science, apart from the standard treatment
to be expected in his commentary on the *Posterior Analytics*.[103] Apparently he was not interested in the calculatory
techniques we have already discussed, nor did he discourse
on the role of mathematical theories in "saving the phenomena," as did Aquinas and others. Zabarella's contribution to

scientific method came rather from the empiricist side, with its accent on observation and experimentation. He frequently employs the term *experientia,* and this not only in his logical writings but in his treatises on physical subjects, and occasionally uses the term *experimentum,* although more in the sense of a thought experiment than in the modern understanding. Charles B. Schmitt has made a detailed comparison of the uses of *experientia* and *experimentum* both by Zabarella and by Galileo in his early writings,[104] and comes to the conclusion that Zabarella was actually the more empiricist of the two. He writes: "What immediately emerges, when one compares Zabarella with the young Galileo, is that, although the latter had recourse to experience more frequently than the former, the very concept of experience has a much more central role in the former's philosophical and scientific methodology." [105] Schmitt does not wish to make extravagant claims for Zabarella, and he readily admits that Galileo "considered a mathematically-oriented approach to be more fruitful than an experientially oriented one," [106] but he does wish to keep the record straight regarding empirical attitudes. Thus he concludes:

> In short, then, Zabarella much more clearly points the way toward seventeenth-century experimental observational science, rooted in the earlier traditions of "magic and experimental science" and of technology, than does Galileo during his pre-Paduan period. If experiment and observation are to be considered the crucial ingredients of "modern science," . . . Zabarella must be considered a much more significant precursor of that movement than the young Galileo.[107]

c. Experimental Method

Other developments in northern Italy contributed to both experimental and mathematical methodologies and thus prepared the way for modern science. An interesting case is the experiment performed by an Averroist Aristotelian and professor at Pisa, Girolamo Borro, who describes it in his *De motu gravium et levium,* published at Florence in 1576.[108] In attempting to resolve an argument as to whether air has weight when in its proper place, he proposes to drop two objects in air, the one having more elemental air in its composition than the other, to see if both fall with the same speed or if the one with the greater air content falls faster.[109] Borro thereupon obtained a small piece of lead and a piece of wood of equal weight, as far as he could judge, and dropped the two simultaneously from a high window. While he and other parties to the dispute watched, contrary to their expectations the wood reached the ground before the lead. This was tried not once but many times (*saepenumero*), always with the same result. From this test (Borro uses *periculum* and *experimentum* interchangeably) he concludes that air must have some gravity in its proper region, since there is more air in the wood than in the lead and the former falls faster through air as a medium.[110] Obviously the results of this experiment are not easy to explain, although Galileo reports a somewhat similar experience and ingenious attempts have been made to interpret it.[111] Our aim here is not to evaluate the experiment but merely to show that an appeal to observational test, whether made properly or not, was regarded by Borro as a definitive way of resolving an argument.

The possibility of performing experiments was also enhanced during the sixteenth century by the rise of a technological tradition largely outside the universities. Machines of increasing sophistication were being designed, both for civil and military purposes, and increasing attention was given to such simple instruments as the balance, the steelyard, and hydrostatic devices for determining weights. These are known in antiquity, of course, but with the more general diffusion of treatises by Archimedes, Hero, and Pappus, a better theoretical understanding of their operation was gained, and improvements consequently made in their construction. Archimedes, in particular, provided a great stimulus to this movement by awakening interest in a mathematical approach to problems of mechanics. The writings of Niccolò Tartaglia, Giovanni Battista Benedetti, and Guido Ubaldo del Monte also opened up new perspectives by way of analyzing the trajectories of cannon-balls, of utilizing specific gravity and centers of gravity when treating problems of dynamics, and of investigating in detail the forces involved in the functioning of balances, levers, and systems of pulleys. Since craftsmen in arsenals, shipyards, and elsewhere increasingly applied these principles to the materials with which they worked, skills were being developed that could be applied to the construction of instruments and experimental apparatus, once the need for empirical tests had finally been realized.[112]

d. Copernicus and Hypotheses

Copernicus, of course, had studied at Padua, and since his theory of the heavens was first presented as merely an attempt to "save the phenomena," it will be fitting to conclude this presentation with a discussion of the status of such hypothetical explanations in the fifteenth and sixteenth centuries. Already in the fourteenth century Pietro d'Abano had opted for Ptolemy's system of eccentrics and epicycles in preference to Aristotle's homocentric spheres on the basis of the argument that "they sufficiently account for the appearances, and do so by the smallest number of motions." [113] The Aristotelians of the fifteenth and sixteenth centuries, on the other hand, reacted strongly against the Ptolemaic system on the ground that it could not be reconciled with physical principles, and generally expressed grave doubts about any system of the universe that proposed merely to "save the appearances."

Achillini, for example, in the first book of his *Quatuor libri de orbibus*, defends the following thesis:

> The motions that Ptolemy assumes are founded upon two hypotheses, the eccentric and the epicycle, which do not agree with physics. Both these hypotheses are false.[114]

He then gives his refutations of these hypotheses and concludes with the statement that Ptolemy's system does not constitute a science but is merely a method for computing entries in astronomical tables. The main burden of his argument is that "astronomers have not established the existence of eccentrics and epicycles by any sort of demonstration," either a priori or a posteriori, "for the effects that are manifest to us may stem from other causes." [115]

Nifo's position is much the same as Achillini's although in this matter, as Duhem has observed, he is influenced more by Aquinas's thought and grants to eccentrics and epicycles the status of a provisional explanation that will suffice until such time as a better one becomes available.[116] Thus he writes:

> You must understand that a good demonstration proves that the cause necessitates the effect, and conversely. Now it is true enough that, the eccentrics and epicycles being conceded, the observed phenomena follow and that they can, therefore, be saved in this way. But the converse does not hold. Starting with the appearances, the existence of eccentrics and epicycles does not follow with necessity; only provisionally, until such time as a better cause be discovered — one which both necessitates the phenomena and is necessitated by them — are the eccentrics and epicycles established. Those therefore err who, starting out from a proposition whose truth may be the outcome of various causes, decide definitively in favor of one of these causes. The appearances can be saved by the sort of hypotheses we have been talking about, but they may also be saved by other suppositions not yet invented.[117]

Nifo, as we have seen, has spoken of three kinds of demonstration, and he explicitly denies that arguments in favor of eccentrics and epicycles belong to any of these categories.[118]

Generally the Renaissance Averroists at Padua were less tolerant in their presentation of astronomical science, and refused to admit any principles that could not be justified in terms of Aristotle's *Physics*. This was the attitude of Girolamo Fracastoro and Gianbattista Amico.[119] Yet other professors at Padua, who may be identified as Ptolemaists, took the opposite position that eccentrics and epicycles are really existent in the heavens and provide the true explanation of the appearances. This was the teaching of Francesco Capuano of Manfredonia, who taught astronomy at Padua and wrote a commentary on Georg Purbach's *Theoricae novae planetarum*.[120] Here he tried to prove that the Ptolemaic hypotheses are not merely sufficient to save the appearances but that they are true and can even be demonstrated. Capuano states:

> Here, as in the *Almagest,* the roads leading to science are the two kinds of demonstration — demonstration by

signs and demonstration in the strict sense. Now the principles of astronomy are inferred a posteriori and from sense: having noted and observed the motion of a planet and the other accidents it presents, one concludes demonstratively, as will be seen from what follows, that this planet has either an eccentric or an epicycle.[121]

It is apparently to this argument that Nifo was objecting in his commentary on the *De caelo*. Others also cautioned against too ready an acceptance of demonstrative proof in such a complex matter. Thus the Dominican Sylvester Mazzolini, in his commentary on Purbach's work,[122] describes the various orbs that have been assigned to the sun by Purbach and Regiomontanus, and evaluates them in this fashion:

> They do not prove that this is the way things are, and perhaps what they assert is not necessary. . . . The sun, then, has three orbs, that is to say, it is believed that the sun has them, but this is not demonstrated; [they] are thought up solely for the purpose of saving what appears in the heavens.[123]

Mazzolini and his Thomistic view, however, represented a minority opinion. And so Duhem rightly concludes his survey of this period with the statement: "Excessive confidence in the reality of the objects involved in the hypotheses of astronomy, or exaggerated distrust of the validity of these hypotheses — these are the two extremes between which the Italian philosophers somehow failed to strike the mean." [124]

Copernicus and his faithful disciple, Joachim Rheticus, both regarded the new theory of the heavens, with its different employment of eccentrics and epicycles, as a true description of reality. Copernicus's approach to astronomy was that of the Italian students with whom he had studied at Padua, in that he intended to save the appearances not merely by a mathematical hypothesis but rather by principles that would be conformable to physics. Well known, of course, is Osiander's preface to the printed edition of Copernicus's *De revolutionibus*, where the opposite impression is created. But scholars who have examined all of the evidence, among them Duhem and Edward Rosen, leave little doubt as to Copernicus's own convictions. As Rosen puts it, Copernicus "was firmly convinced that he was talking about the actual physical world when he transformed the earth from the sluggish dregs of the universe to a satellite spinning about its axis as it whirled around the sun." [125]

Finally, behind much of fifteenth- and sixteenth-century thought about the universe were ideas quite different from those current in the thirteenth and fourteenth centuries, although having something in common with the views of Grosseteste with which we started our account of medieval methodology. Renaissance humanism saw a return to the Greek text of Plato as well as to that of Aristotle, and thus Platonic and Neoplatonic concepts such as the world soul (*anima mundi*) again came into fashion. Not only was there the Neoplatonic accent on geometry, and even the attribution of special and secret significance to mathematical insights in the Neopythagorean mode, but there was also a more general acceptance of explanations in terms of spirit and occult qualities. Nicholas of Cusa, for example, held that impetus animates the heavenly spheres just as the soul animates the human body, and even used this analogy to explain the spinning of a top. Giordano Bruno, as is well known, held that the spheres of the universe were endowed with life and soul, and Marsilio Ficino stressed such principles as that of cosmic sympathy, of "like attracting like," to explain the motions of the universe. Similar ideas are to be found in Copernicus's explanation of gravity and Fracastoro's treatment of the magnet. By the end of the sixteenth century, therefore, the way was prepared for the teachings of Gilbert and Kepler, and animistic notions that had been rejected by the Averroist and scholastic traditions, and by Aquinas in particular, came even to be identified as "peripatetic." [126]

In this complex, if not confusing, fashion was the stage set for the "new science" of the seventeenth century. True to their medieval heritage, investigators of nature were convinced of the existence of a real world and had confidence in the ability of empirical and mathematical reasoning to uncover the causes of physical phenomena. Hypothetical explanations were countenanced in some quarters, but generally the ideal of science was that set by the *Posterior Analytics* of Aristotle. Little progress, it is true, had been made in laying bare the secrets of nature, but this could be attributed more to lack of skill in application than to a defect in the methods available. And thus there seemed to be, at the close of the Renaissance, despite the continued presence of numerous and vexing substantive problems, a general feeling that the methodological canons were well in hand. Mankind was entering a new and exciting era of discovery, as was immediately apparent in the geographical sphere, and one had little reason to fear that the progress in evidence would not soon extend to the scientific sphere as well.

Part Two
Early Classical Science

CHAPTER FIVE

The Founders of Classical Science

DURING the discussions of the third day in the *Discorsi*, after having explained his new science of motion and the law of falling bodies, Galileo has Sagredo raise a key question: "From these considerations it appears to me that we may obtain a proper solution of the problem discussed by philosophers, namely, what causes the acceleration in the natural motion of heavy bodies?" [1] A brief discussion follows, and then Salviati gives the evasive, if not agnostic, answer: "The present does not seem to be the proper time to investigate the cause of the acceleration of natural motion . . ." [2] The reason he offers is simple enough: there has been so much discussion of this problem by so many philosophers who have offered so many conflicting explanations that any further examination "is not really worth while." [3] And so Salviati concludes that "at present, it is the purpose of our author [Galileo] merely to investigate and to demonstrate some of the properties of accelerated motion, whatever the cause of this acceleration may be . . ." [4]

Writing almost eighty years later, in the General Scholium he appended to the second edition of his *Mathematical Principles of Natural Philosophy*, Sir Isaac Newton sums up all of his previous work on a similar note: "Hitherto we have explained the phenomena of the heavens and of our sea by the power of gravity, but have not yet assigned the cause of this power."[5] He goes on to explain what he thinks he has proved with regard to gravitation and its various properties, and again admits, "I have not been able to discover the cause of those properties of gravity from phenomena, and I feign no hypotheses . . ."[6] In other words, his search for causes seemingly has come to an impasse, and like Galileo he must settle for the laws governing phenomena and remain agnostic with regard to the underlying causes that produce them.

From these statements of the most celebrated founders of modern science, Galileo and Newton, one might conclude that the classical period of modern science, say from the seventeenth to the nineteenth centuries, saw a radical departure from the search for causal explanations we have presented as typical of the medieval and Renaissance periods. One might thus be tempted to say that classical scientists gave up the quest for answers to the question "Why?" and were henceforth content to settle for answers to the question "How?" Such a characterization, while offering the advantages of simplicity, unfortunately would not be adequate for understanding the methodological innovations that attended the appearance of the "new science" and its growth and gradual perfection during the subsequent centuries.

In the era we now begin to discuss, which stretches from William Gilbert (b. 1544) to Claude Bernard (d. 1878), a diversity of methodological convictions, and in particular a diversity of attitudes toward causality, came into evidence. At the outset of the period the tradition of the *Posterior Analytics* was still strongly influential, so much so that one could say that the founders of modern science, whatever their protestations against Aristotle, were for the most part captive to his methodological search for causes. When they rejected the peripatetic tradition, this was usually on the supposition that the peripatetics had been concerned with occult or fictitious causes, whereas in the light of their own efforts the "true causes" of natural phenomena would now become manifest. As we have just seen in the citations from Galileo and Newton, of course, even this search for "true causes" ran into impasses, and so philosophers who were stimulated by the scientific enterprise began to question the knowability of causes, first those of the Aristotelian type and ultimately the "mechanical causes" that had come to replace them. While such a skeptical or agnostic attitude continued to develop and be refined through the eighteenth and nineteenth centuries, however, this same period saw repeated flirtations with rationalism and empiricism in efforts to provide an epistemology, if not an ontology, for the new science. And strangely enough, both rationalists and empiricists continued to speak of causality, the former usually meaning by this a knowledge of forms, themselves intelligible and really laws of nature that governed phenomena from within, and the latter meaning the motions of atomic particles that would serve to explain the temporal sequence of observable events. Thus it was that those who concerned themselves *ex professo* with scientific methodology were loathe to give up causal terminology, although they themselves brought about an evolution in the meaning of its terms. By the end of the nineteenth century, for example, under their influence causal explanation had come to be identified in the minds of scientists with determinism and predictability, and so became itself vulnerable when determinism and prediction were no longer to be regarded as attainable in physical science.

To locate this complex development in some kind of chronological sequence, it will be necessary to anticipate the division of subject matter to be treated in the next volume. We propose to cover the period of classical science in three chapters, the first of which will bring our exposition of medieval and Renaissance science in this volume to a close; the latter two chapters, as more fittingly introducing the contemporary problematic, will be found in the second volume. Thematically, the three chapters treat successively the contributions of the founders of classical science, of the philosophers who were influenced by them, and of the methodologists who codified the final results. This thematic division, of course, poses a problem as to which thinkers are to be discussed in each chapter, particularly when a chronological development is also being traced. Some figures, such as Descartes and Leibniz, and to a lesser extent Francis Bacon, could easily qualify for inclusion in more than one category, and thus a degree of arbitrariness is inevitable in the selection of those to be treated in each. To keep such arbitrariness to a minimum it is proposed to treat individuals in the group for which their contributions have proved most significant; on this basis Descartes and Leibniz fall to the philosophers and Bacon to the methodologists. The founders of classical science, applying the same rule, can be restricted to Gilbert, Kepler, Galileo, Harvey, and Newton; among the philosophers, apart from Descartes and Leibniz, will then be found Hobbes, Locke, Berkeley, Hume, and Kant; and the methodologists, apart from Bacon, will include Auguste Comte, John Herschel, William Whewell, John Stuart Mill, and Claude Bernard. Predominant attention will thus be paid, successively, to the founders of classical science in the seventeenth century, to its philosophers in the seventeenth and eighteenth, and to its methodologists in the nineteenth.

Throughout these three chapters the treatment of causal explanation will be for the most part sympathetic, noting instances of its use and diversities of its interpretation, but generally refraining from judgment or criticism in the light of contemporary developments in the philosophy of science, which will be taken up in the latter part of the next volume. Again, because of the sheer extent of the materials to be discussed, the treatment must be selective, even more so than in the preceding chapters of this volume. Thus we begin in this chapter with only those practitioners who made classical contributions and offered explanations that were to become paradigmatic for future developments. These include Gilbert and Kepler for their work in magnetism and astronomy, Galileo and Harvey as the fathers of physical and biological science respectively, and Newton as the synthesizer who, more than any other, created the classical conception of science that was to hold sway until practically the end of the nineteenth century.

1. *William Gilbert*

William Gilbert studied at Cambridge in both the arts faculty and the faculty of medicine, and achieved renown as a physician before writing his celebrated treatise on the magnet.[7] He composed also an unfinished work on cosmology, *De mundo nostro sublunari philosophia nova,* which was not published until half a century after his death and had little influence on that account. Himself an experimenter of genius, he frequently called attention to his "experiments and discoveries," persuading his readers to repeat them for themselves. Apart from his experimentalism, Gilbert is noteworthy for his attempt to explain gravitational phenomena as a type of magnetic attraction, for his establishing a distinction between magnetic and electric phenomena as pertaining to different fields of study, and for his introduction of the term "electric" into the English scientific vocabulary.

Although the six books that make up the *De magnete* contain a wealth of descriptive and experimental data, their entire orientation is toward the discovery of the causes of magnetic phenomena. Gilbert was aware that others had noted the types of movement performed by magnets, but in general he is critical of their lack of success in discovering the "true causes" [8] of such phenomena. His own outstanding discovery is that "magnetic bodies are governed and regulated by the earth, and they are subject to the earth in all their movements." [9] So he writes:

> All the movements of the loadstone are in accord with the geometry and form of the earth and are strictly controlled thereby, as will later be proved by conclusive experiments and diagrams; and the greater part of the visible earth is also magnetic, and has magnetic movements, though it is defaced by all sorts of waste matter and by no end of transformations. [10]

In order to make precise the type of causality that the earth exercises in magnetic movements, Gilbert found it necessary to make a distinction between electric and magnetic phenomena. As he diagnosed the source of the difference between the two, "in all bodies everywhere are presented two causes or principles whereby the bodies are produced, to wit, matter (*materia*) and form (*forma*). Electrical movements come from the *materia,* but magnetic from the prime *forma;* . . ." [11] Both types of phenomena give evidence of a sensible attractive influence, since the "two kinds of bodies . . . are seen to attract bodies by motions perceptible to our senses":

Electrical bodies do this by way of natural effluvia from humour; magnetic bodies by formal efficiencies or rather by primary native strength (*vigor*). This form is unique and peculiar: it is not what the Peripatetics call *causa formalis* and *causa specifica in mixtis* and *secunda forma;* nor is it *causa propagatrix generantium corporum;* but it is the form of the prime and principal globes; and it is of the homogeneous and not altered parts thereof; the proper entity and existence which we may call the primary, radical, and astral form; not Aristotle's prime form, but that unique form which keeps and orders its own globe. Such a form is in each globe — the sun, the moon, the stars — one; in earth also 'tis one, and it is that true magnetic potency which we call the primal energy. Hence the magnetic nature is proper to the earth and is implanted in all its real parts according to a primal and admirable proportion.[12]

Gilbert knew, of course, of Peter of Maricourt's letter on the magnet and was greatly influenced by it. From Peter he apparently got the ideas that the loadstone or magnet was naturally spherical in shape, that the world itself was a giant magnet, and that a spherical loadstone could be regarded as a *terrella*, or little Earth. From these notions it was a simple step to conclude that every magnet on the earth's surface should behave just like the iron filament placed on Peter's loadstone globe. Peter had thought, moreover, that magnets derive their power not only from the poles but from the heavens as a whole; Gilbert rejects this explanation, and instead places the causality directly in the form associated with earth. He explains:

It [the magnet's power] is not derived from the heavens as a whole, neither is it generated thereby through sympathy, or influence, or other occult qualities: neither is it derived from any special star; for there is in the earth a magnetic strength or energy of its own, as sun and moon have each its own *forma;* and a little fragment of the moon arranges itself, in accordance with lunar laws, so as to conform to the moon's contour and form, or a fragment of the sun to the contour and form of the sun, just as a loadstone does to the earth or to another loadstone, tending naturally toward it and soliciting it.[13]

Having established his principle of magnetic explanation in this way, Gilbert explains how matter is of its nature limited whereas form is not,[14] and then discourses on all of the effects of *forma* in producing magnetic phenomena. Among such effects he enumerates the directive force that orients the magnet, the variation or declination from true north, and the dip of a magnetic needle on the earth's surface.

Of these the first is "a force distributed by the innate energy from the equator in both directions to the poles"; it accounts for a magnetic needle's movement to the north-south direction, and also for its stability or "constant and permanent station in the system of nature." [15]

More mysterious and difficult to explain is the second phenomenon of declination, which in Gilbert's view has an "occult and hidden cause"; [16] all of Book IV, in fact, is devoted to uncovering such a cause and supplying a demonstration for it. Gilbert begins his treatment of this phenomenon, which he called "variation," with the following preamble:

> So far we have been treating of direction as if there were no such thing as variation; for we chose to have variation left out and disregarded in the foregoing natural history, just as if in a perfect and absolutely spherical terrestrial globe variation could not exist. But inasmuch as the magnetic direction of the earth, through some fault or flaw, does depart from the right track and the meridian, the occult and hidden cause of variance which has troubled and tormented, but to none effect, the minds of many has to be brought to light by us and demonstrated. They who hitherto have written of the magnetic movement have recognized no difference between direction and variation, but hold there is one only movement of the magnetized needle.[17]

Gilbert, while recognizing this hitherto overlooked distinction between orientation and declination, then attributes both phenomena to a common cause, the earth:

> As we have already said, every movement of loadstone and needle, every turn and dip, and their standing still, are effects of the magnetic bodies themselves and of the earth, mother of all, which is the fount and source and producer of all these forces and properties. Thus, then,

the earth is the cause of this variation and tendence to a
different point in the horizon; but we have to inquire further how and by what potencies it acts.[18]

Gilbert thereupon rejects the explanations that have been offered by others, and gives his own in terms of the fact that
the "globe of the earth is at its surface broken and uneven,
marred by matters of diverse nature . . ."[19] It is therefore
"the earth, by reason of lateral elevations of the more ener-
getic globe, [that] causes iron and loadstone to diverge a few
degrees from the true pole or true meridian."[20]

The phenomenon of magnetic dip is the next topic investigated, and most of Book V is devoted to its causal explanation. Here Gilbert is confident that he has discovered "the true
and definite cause of this great and hitherto unknown
effect"[21] and writes about it in a style that will become quite
common among seventeenth-century scientific authors:

> We come at last to that fine experiment, that wonderful
> movement of magnetic bodies as they dip beneath the
> horizon in virtue of their natural verticity; after we have
> mastered this, the wondrous combination, harmony, and
> concordant interaction of the earth and the loadstone (or
> magnetized iron), being made manifest by our theory,
> stand revealed. This motion we have so illustrated and
> demonstrated with many experiments, and purpose in
> what follows so to point out the causes and reasons, that
> no one endowed with reason and intelligence may justly
> contemn, or refute, or dispute our chief magnetic princi-
> ples.[22]

Then, having recounted all these principles and the experiments used to demonstrate them, Gilbert feels impelled to
generalize his explanation and show how "the formal magnetic act [is] spherically effused":[23]

> Repeatedly we have spoken of the poles of earth and ter-
> rella and of the equinoctial circle; last we treated of the
> dip of magnetized bodies eastward and terrellaward, and
> the causes thereof. But having with divers and manifold
> contrivances laboured long and hard to get at the cause
> of this dip, we have by good fortune discovered a new
> and admirable science of the spheres themselves — a sci-
> ence surpassing the marvels of all the virtues magnetical.

For such is the property of magnetic spheres that their force is poured forth and diffused beyond their superficies spherically, the form being exalted above the bounds of corporeal nature; and the mind that has diligently studied this natural philosophy will discover the definite causes of the movements and revolutions.[24]

Unfortunately these wonderful powers of the earth's form led Gilbert to speculate further that the magnetic force is "as it were, animate,"[25] and that it "imitates a soul,"[26] thereby adapting the Neoplatonic world-soul doctrine to scientific applications. Much impressed with Copernicus's treatise on the revolution of the celestial spheres,[27] Gilbert wished even to account for the earth's daily rotation with this magnetic force and the functioning of the earth's "soul."[28] As he explains it:

By the wonderful wisdom of the Creator, therefore, forces were implanted in the earth, forces primarily animate, to the end the globe might, with steadfastness, take direction, and that the poles might be opposite, so that on them, as at the extremities of an axis, the movement of diurnal rotation might be performed.[29]

The earth's motion and the magnetic forces producing it, moreover, would be influenced by the sun, and this facet of Gilbert's teaching would tie him in with Copernicus's heliocentrism, as it was later to stimulate Kepler to a similar type of explanation:

The earth therefore rotates, and by a certain law of necessity, and by an energy that is innate, manifest, conspicuous, revolves in a circle toward the sun; through this motion it shares in the solar energies and influences; and its verticity holds it in this motion lest it stray into every region of the sky. The sun (chief inciter of action in nature), as he causes the planets to advance in their courses, so, too, doth bring about this revolution of the globe by sending forth the energies of his spheres — his light being effused.[30]

According to Gilbert, therefore, "the sun itself is the mover and inciter of the universe,"[31] and "the causes of the diurnal motion are to be found in the magnetic energy,"[32] which is apportioned out in various ratios so as to keep the entire universe revolving "symmetrically with a certain mutual concert and harmony." [33] So ambitious was Gilbert to thus account for the then known astronomical phenomena in terms of magnetic force that he devoted two chapters at the end of his treatise to magnetic explanations of the precession of the equinoxes. Finally having to admit defeat in this last enterprise, he concludes with the following observations: "Hence all these points touching the unequal movement of precession and obliquity are undecided and undefined, and so we cannot assign with certainty any natural causes for the motion." [34] But even this admission confirms his consistent search for causes, not only of magnetic phenomena but of all the observable movements of the earth and the stars.

2. Johannes Kepler

Johannes Kepler [35] knew of Gilbert's work and was much impressed by it. Himself a mathematician and astronomer of ability, he was likewise attracted to the Neoplatonic type of philosophy, and this even more so than Gilbert. He continued the latter's search for causes, particularly in his attempts to explain the system of the universe; in so doing, like Gilbert, he sought formal causes in the mathematical or Neopythagorean mode, but at the same time he was enough of an Aristotelian to search for the "natural" or efficient causes that might also serve to explain the phenomena of the heavens. So, in effect, Kepler attempted a new synthesis of Neoplatonic and Aristotelian thought. He was acquainted with the writings of Nicholas of Cusa and of Giordano Bruno, and in a letter to Galileo referred to Plato and Pythagoras as "our true masters." [36] At the same time, in his *Epitome of Copernican Astronomy* he made use of Aristotle's *Posterior Analytics*, *Physics*, and *Metaphysics*, and even proposed Book 4 of the *Epitome* as "a supplement to Aristotle's *De caelo*." [37]

Apart from these classical influences, Kepler was an admirer of Copernicus, whose system of the world he had been taught by Maestlin while studying at Tübingen, and which appealed to him because of its simplicity and mathematical harmonies.[38] Tycho Brahe appealed to him also because of the precision of his astronomical observations, which Kepler recognized as a necessary starting point for his own calculations.[39] Finally he was indebted to Gilbert for likening gravitational phenomena to those of magnetic attraction, which suggested a deeper insight into the physical causes that might be able to explain the mathematical harmonies of the universe. Quite truthfully could he therefore affirm in his *Epitome:* "I build my whole astronomy upon Copernicus' hypotheses concerning the world, upon the observations of Tycho Brahe, and lastly upon the Englishman, William Gilbert's philosophy of magnetism." [40]

Kepler's dependence upon Brahe's observational astronomy gives clear indication that one of his major purposes was "to save the phenomena" of the heavens. He was not content, however, to stop there, but wished also "to contemplate the true form of the edifice of the world." [41] How he proposed to go from the appearances to the underlying reality of this "true form" is not completely clear, but there is little doubt that Kepler's theological views suggested a possible method. He was convinced that God, "like a human architect," had created the universe "according to order and rule and measure," [42] and therefore that the universe was constructed on the model of archetypal ideas existing in the divine mind.[43] The quantity and harmony of the universe found ready expression, for him, in Plato's doctrine of ideas, and he had only to combine these with some elements of Christian revelation to be convinced of their explanatory force in accounting for the physical universe. For example, following an Augustinian line of thought, Kepler believed that the Trinity was mirrored in the visible universe, and that a spherical universe, particularly one with the sun at the center, was a symbolic representation of the Trinity. As he put it, "The image of the triune God is in the spherical surface, that is to say, the Father is in the center, the Son is in the outer surface, and the Holy Spirit is in the equality of relation between point and circumference." [44]

The particular way in which Kepler understood the sun as the symbol of the Father is of importance for understanding his synthesis of Platonism and Aristotelianism. According to Plato's account in the *Timaeus* the world was made "in the form of a globe, round as from a lathe, having its extremes in every direction equidistant from the center, the most perfect and most like itself of all figures . . ." [45] When the Demiurge fashioned the universe, moreover, "in the center he put the soul, which he diffused throughout the body." [46] Thus, for Plato, the Demiurge's animating force radiated from the center of the universe outward to its periphery. For Aristotle, on the other hand, the Prime Mover was not located at the universe's center, but rather was a motive force that was applied to the outermost sphere and kept all of the heavenly spheres revolving, while exerting its causality also in the sublunary sphere. Kepler, like Aristotle, saw the need for a continued application of force to sustain motion, but he preferred to locate its origin in the center of the universe rather than at its periphery. Moreover, since the sun is the symbol of God the Father, it must be located, as Copernicus had proposed, at the center of the world, where it is the source of light and heat as well as the generator of the forces that drive the planets in their circumsolar orbits. These arguments, while symbolic in nature, seemed to receive strong confirmation from the fact that a sun-centered universe is geometrically simpler than an earth-centered universe, particularly since no major epicycle would be needed to explain the retrograde motion of the planets in the heliocentric arrangement.

This general type of explanation, viewed in the context of Aristotle's four-fold causality, places heavy stress on formal causality, although it also allows for the exercise of efficient causality, and even recognizes the need for a material cause or substrate and for a basic teleology or final causality in the operation of the universe. Kepler concentrated, however, on the formal and efficient causes, since in his view "the natural and archetypal causes of celestial physics" go together.[47] The geometrical harmony of the universe is accounted for by archetypal causes, whereas the dynamic harmony is explicable in terms of natural or efficient causes. Kepler was probably led to emphasize these two types of explanation, as Burtt has argued, because his mathematicism, which he had inherited from Nicholas of Cusa and others, was tempered by an empiricism that grew from his associations with Brahe.[48] On this interpretation, although the geometrical arrangement of the universe, which served for Kepler as the "Aristotelian formal

cause reinterpreted in terms of exact mathematics,"[49] could
provide a sufficient answer to the problem of the number and
size of planetary orbits, it could not account for the varying
speeds of the planets in their orbits, which had been empirically established by Brahe. It was thus the desire to solve the
dynamical problem presented by such variation of speed that
induced Kepler to investigate the efficient causes operative in
the world. And it was here that he was able to take inspiration from Gilbert, and from the analogy the latter had dis-
cerned between gravitational and magnetic attraction.

This line of development in Kepler's thought is evident
in a letter written by him to David Fabricius on November
10, 1608, where he attempts to answer Brahe's argument that
the earth does not rotate because bodies thrown upward from
the earth's surface always return to the same point. Kepler's
answer is that the projected body is kept moving along with
the earth by invisible magnetic lines or chains. So he queries:

> How is it possible that a sphere, thrown vertically
> upward — while the earth rotates meanwhile — does re-
> turn to the same place? The answer is that not only the
> earth, but together with the earth, the magnetic invisible
> chains rotate by which the stone is attached to the un-
> derlying and neighboring parts of the earth and by
> which it is retained to the earth by the shortest, that is,
> the vertical line.[50]

In the introduction to his *Astronomia nova*, written only a
year later, Kepler elaborates on the concept of gravity, and,
while continuing to compare it to a "magnetic faculty" with
an "animating" function, stresses that it is a "mutual affec-
tion" emanating from bodies. This time he asserts categori-
cally:

> Here is the true doctrine of gravity. Gravity is a mutual
> affection among related bodies which tends to unite and
> conjoin them (of which kind also the magnetic faculty
> is), while the earth attracts the stone rather than the
> stone tends towards the earth. Even if we placed the cen-
> ter of the earth at the center of the world, it would not
> be toward the center of the world as such that heavy
> bodies would be carried, but rather toward the center of

the round body to which they are related, that is, toward the center of the earth. Thus, no matter whereto the earth is transported, it is always toward it that heavy bodies are carried, thanks to the faculty animating it.[51]

An attraction of this Keplerian type is always exerted toward bodies, since a mathematical point has no power to move a heavy body.[52] Here Kepler is merely echoing something that Gilbert had stated in his *De magnete*, namely, that it is "in bodies themselves" that the acting force resides, "not in space" or "interspaces."[53] Although mathematical points do not attract, Kepler nonetheless envisages that any two bodies sufficiently isolated from a third body will mutually attract each other so that they will come together at some point intermediate between the two. Thus he maintains:

If two stones were removed to some place in the universe, in propinquity to each other, but outside the sphere of force of a third cognate body, the two stones, like magnetic bodies, would come together at some intermediate place, each approaching the other through a distance proportional to the mass (*moles*) of the other.[54]

All of these mutual affections and the effects they produce are, moreover, for Kepler susceptible to mathematical treatment, since they are basically spatial concepts and therefore follow mathematical laws. So he argues:

For we see that these motions take place in space and time and this virtue emanates and diffuses through the spaces of the universe, which are all mathematical conceptions. From this it follows that this virtue is subject also to other mathematical "necessities."[55]

This account of gravitational force, already present in Kepler's early writings, provides the necessary background for an understanding of his fuller treatment of celestial mechanics in the *Epitome*. Utilizing Brahe's extremely accurate observations of planetary positions, Kepler became aware of the fact that the planets do not travel with uniform speed around their orbits, but move more rapidly the nearer they approach the sun. Kepler's first formulation of this, later to become his famous "second law," was that a planet's velocity of travel is always inversely proportional to its distance from the sun. Speculating on the reason for this in his *Astronomia nova*, Kepler allows a series of possibilities: either the decrease of velocity is the cause of the increase of distance from the sun; or the increase of distance is the cause of the decrease of velocity; or it is possible that both are reducible to a common cause, namely, a decrease of the force acting on the planet as it moves farther and farther from the sun. As is known from his other writings, Kepler thought of the planets as magnets driven around the sun by a type of magnetic force.[56] Consistent with this understanding, his view of gravity was not that it exerted an attractive force, but rather that its influence was one of propelling the planets along their orbits.[57] Such a view of motor causality was quite in accord with the Aristotelian principle that "omne quod movetur ab alio movetur," since, as we have already noted, Kepler regarded the continued application of a mover as necessary to sustain motion.

Kepler's early understanding of gravitational force, based on the analogy of magnetic attraction, was expressed in animistic terms. While it is true that, even in his later writings, he continued to think of the universe as energized by a type of soul, his earlier conception nonetheless underwent a gradual development. He explains this in his annotations to the second edition of the *Mysterium cosmographicum* (1621):

> If you substitute for the word "soul" the word "force," you have the very principle on which the celestial physics of the treatise on Mars etc. is based . . . Formerly I believed that the cause of the planetary motion is a soul, fascinated as I was by the teachings of J. C. Scaliger on the motory intelligences. But when I realized that these motive causes attenuate with the distance from the sun, I came to the conclusion that this force is something corporeal, if not so properly, at least in a certain sense.[58]

Having made the acknowledged transition from a soul principle to the concept of a central force localized in a body at the center of the world, i.e., in the sun, Kepler was still faced with two problems: (1) how to explain the non-uniform motion of the planets, and (2) how to explain the fact, which he himself had previously discovered, that the planets do not move in perfect circles but rather in oval or elliptical orbits. If a single force alone were acting on the planets, it would seem that each planet should undergo a uniform or regular movement in a perfect circle; but this is not borne out by empirical observation. Since, in Kepler's mind, causes or forces must also be operative to produce the experienced non-uniformity, he was finally led to propose that each planet must be subjected to two conflicting influences: one a force emanating from the sun and accounting for the planet's orbital motion, the other a force located in the planet itself and accounting for that motion's irregularities.[59]

Such an explanation is obviously based on reasoning along lines of efficient causality, but for Kepler it had a deeper root in archetypal or formal causality, since he tended to see natural or efficient causes as subservient to a higher archetypal cause. The way he justified such an insight may be seen from his calculations with respect to the motion of the earth around the sun. If the earth followed its primary archetype it should make 360 daily revolutions in the course of the year. As a matter of fact, however, it exceeds this archetypal number by 5¼ additional revolutions, and so another cause must account for this. "For if the number 365¼ were not composed of the two effects of two distinct causes, there would be no reason why it is not one of the archetypal numbers." [60] And again: "If the sun caused everything, the whole diurnal revolutions of the earth would be proportional to the intervals between the sun and the earth." [61] It is this type of reasoning that led Kepler to introduce, apart from the sun's force, the planet's inertia, or laziness, as the perturbing factor that would produce both non-uniform velocities and elliptical orbits. The resulting interaction of the planet with the sun he describes as follows:

> Not only do the motor virtue and the movable body come together in the movements, but also the inward rectilinear configuration of the movable body; and in proportion to its diversity of posture in relation to the sun, this configuration is affected in diverse ways in the movement; in one region it is repelled, in another it is attracted toward the inside. . . . [Ultimately] the figure of its route becomes elliptical.[62]

In other texts Kepler explains how this interaction is the result of two component forces: the one is a central force that propels the planet in its circumsolar orbit, whereby the sun, having "laid hold of the planet, either attracting it or repelling it, or hesitating between the two, makes the planet also revolve with it;" [63] the other is "a natural inertia" whereby the planet reacts to the sun's influence, with the result that "the motor power of the sun and the powerless or material inertia of the planet are at war with one another." [64] It is the tension resulting from the conflict of these two forces that ultimately accounts for the non-circular and non-uniform motion of the planets. And in yet other texts Kepler attempts to relate this motive force of the sun with the luminous energy that proceeds from that body, showing how there are some similarities between them but at the same time that they are distinct. The reason is that:

> Light is bounded and stopped by the surfaces of opaque bodies, so that it goes on no farther into other bodies lying in the same straight line. But this force which moves the planet by laying hold of it is not stopped by its surface, but goes into the body which it lays hold of, and moreover goes on through the body into the body of a farther planet, if it so happens that two planets are on a straight line with the sun . . .[65]

While pioneering with these force concepts, which in many ways anticipated Newton's treatment of planetary motion, Kepler himself was disappointed that he had to introduce the notion of elliptical orbits and could not explain planetary motions in terms of perfect archetypal circles. As a mathematician he was principally interested in harmonic ratios and in detecting the hidden regularities in the universe, so much so that he experienced great joy in discovering his third law, which affirms a simple geometrical relationship between the periods of planets and their distances from the sun. It was this continual search for the underlying geometry of the universe, which may be regarded as a search for formal causes, that most delighted Kepler. But at the same time he was enough of an empiricist, in the sense of being constrained to account for the phenomena as these were actually observed, to know that he had to introduce efficient causes when dealing with the inertial type of matter that seemed to be involved in planetary motion. It is for this latter reason, especially, that Max Jammer sees Kepler as marking the decisive stage in the introduction of the concept of force into the exact sciences. Jammer recognizes a variety of circumstantial reasons that impelled Kepler to introduce the force concept, but in his analysis these were all overshadowed by a more technical and methodological reason, namely, Kepler's "desire for causal explanation." [66]

3. Galileo Galilei

Galileo is almost universally regarded as the key figure in the foundation of modern science, even more than Gilbert and Kepler, and yet he presents an enigmatic figure when one wishes to explain his attitude toward causality and its role in scientific explanation. Part of the difficulty stems from Galileo's intellectual development as he passed through various stages at Pisa, Padua, and Florence. Again, the nature of the work he was called upon to do, and the controversies in which he became engaged, make it difficult to disengage his true views from the polemics in which they were frequently immersed. Although he does make statements that would lead one to believe he held an agnostic, if not a positivist, attitude toward causality, there is other evidence to show that he likewise sought causal explanations, and regarded these as the ultimate to which science could attain.[67]

Galileo was a contemporary of Kepler and Gilbert and knew of their teachings, although he disagreed with particular points of their causal interpretations of nature. He received his early intellectual formation at the University of Pisa, where he also taught for several years, and then moved to the University of Padua for a more prolonged teaching career. During his early years at Pisa he gives evidence of having studied the Aristotelian tradition, as witnessed by his student notebooks, edited by Favaro under the title of *Juvenilia*,[68] and also from his *Quaestiones* on logic, which discuss in considerable detail questions similar to those raised by Nifo and Zabarella relating to Aristotle's *Posterior Analytics*.[69] These writings reveal an understanding of, and a general commitment to, Aristotle's teaching on causality and its role in demonstrative proof. In fact, when organizing the material of the *Juvenilia,* Galileo not infrequently makes his divisions on the basis of the four causes; here he follows a pattern used also by one of his teachers, Francesco Buonamici, in his lengthy treatise entitled *De motu libri decem.*[70] Some have argued on such grounds that the *Juvenilia* are actually notes derived from Buonamici's lectures or on those of another of Galileo's professors while he studied at Pisa. On the other hand, as we have shown elsewhere, there are enough evidences of originality in Galileo's composition to question that his work is merely derivative and to suggest the possibility that in the early stages of his intellectual life he subscribed to the main Aristotelian theses, and in fact interpreted them along Thomistic lines, although with a certain element of eclecticism reflecting Scotistic and Averroistic influences.[71]

a. The Early *De motu*

In his first treatise on mechanics, entitled *De motu,* however, there can be no doubt that Galileo had changed his intellectual stance and was arguing in a pronounced anti-Aristotelian vein.[72] No less than six chapters of this work begin with the announcement that the conclusion is contrary to Aristotle's, and in one place Galileo declares outright: "Aristotle, in practically everything that he wrote about local motion, wrote the opposite of the truth."[73] Such a change of attitude from the *Juvenilia* to the *De motu* is not too difficult to explain, as has been pointed out, when one considers that the *De motu* was written while Galileo was a professor of mathematics at Pisa.[74] In this capacity he was quite fascinated with Euclid and Archimedes, and he was not bound to any doctrinal loyalty in matters of natural philosophy. As a mathematician, moreover, he was much interested in the problems of local motion, which he could rightfully feel had received perfunctory treatment at the hands of the professors who had taught him. His own acknowledged master was the "divine,"[75] the "superhuman Archimedes, whose name I never mention without a feeling of awe,"[76] and his criticism of Aristotle, significantly, was that he relied too heavily on experience and so was "ignorant not only of the profound and more abstruse discoveries of geometry, but even of the most elementary principles of this science." [77]

Granted the anti-Aristotelian polemic obvious in the *De motu,* which was to continue through most of his later writing, one may inquire whether the methodology Galileo employed in the *De motu* is significantly different from that of the *Posterior Analytics,* or, more pointedly, whether at this stage of his investigations Galileo had abandoned the search for the causes of local motion. The answer to both questions can only be in the negative. As several Galileo scholars have pointed out, it was not the method of the *Posterior Analytics* that Galileo questioned but rather its use by the contemporary Aristotelians, whom he felt were superficial and relied too much on authoritative argument, rather than themselves searching for the causes of the effects they observed.[78] Again, when attempting to evaluate Galileo's early version of the *De motu* (and the other materials gathered under this title by Favaro), one sees immediately that they are not too different from Buonamici's *De motu,* or from Zabarella's *De motu gravium et levium,* or even from Girolamo Borro's work of the same title.[79] All are essentially Aristotelian treatises in that they analyze projectile motion and gravitational motion in terms of the causes that produce them.

By way of illustration of this fact, Chapter 7 of Galileo's treatise inquires into "the cause of speed and slowness of natural motion." [80] The title of Chapter 10 reads "In which, in opposition to Aristotle, it is proved that if there were a void motion in it would not take place instantaneously but in time." [81] Admittedly the latter is an anti-Aristotelian teaching, but no more so than that of Thomas Aquinas or Domingo de Soto, both of whom disagreed with Aristotle on this matter, though on a basis different from Galileo's.[82] Chapter 17 raises the question of the efficient cause of the motion of projectiles, and here Galileo again rejects Aristotle's conclusion, which was that the projectile was moved by a disturbance produced, by the projector, in the medium through which the projectile passes. Galileo's own account is comparable to Aristotle's:

Let us therefore conclude finally that projectiles can in no way be moved by the medium but only by a motive force impressed by the projector. And let us now go on to show that this force is gradually diminished and that in the case of forced motion no two points can be assigned in which the motive force is the same.[83]

Here effectively Galileo has adopted Buonamici's solution to the projectile problem, while still analyzing it in terms of efficient causality, as was the practice among contemporary Aristotelians.

As another example, in Chapter 19 Galileo takes up a question "In which the cause of the acceleration of natural motion towards its end is set forth, a cause far different from that which the Aristotelians assign." [84] Here he is explicit that he is searching for a causal explanation, which he recognizes as being quite difficult to uncover, but which he nevertheless believes he has succeeded in doing:

> The reason why the speed of natural motion is increased toward the end is certainly more difficult to discover than to explain. Either no one has thus discovered it or, if at times some one has hinted at it, he has presented it in imperfect and defective form and it has not been accepted by philosophers in general. While engaged in seeking the cause of this effect [of acceleration] which I shall not call surprising but necessary — for the cause assigned by Aristotle never appealed to me — I was troubled for a long time and did not find anything that fully satisfied me. And, indeed, when I discovered an explanation that was completely sound (at least in my own judgment), at first I rejoiced. But when I examined it more carefully I mistrusted its apparent freedom from any difficulty. And now finally, having ironed out every difficulty with the passage of time, I shall publish it in its exact and fully proved form.[85]

Following this Galileo enters into a detailed discussion of Aristotle's explanation and then presents his own solution. Briefly, he argues that at the beginning of a body's descent there is a considerable force that impels it upward, which diminishes the essential weight of the body and makes its motion slower at the beginning. As the body continues to fall, however, the retarding effect of the lightness is weakened and the effective weight of the body becomes greater, causing the body to move faster and faster. Galileo concludes his exposition with the assuring words, "This is what I consider to be the true cause of the acceleration of motion." [86] Note here the expression "true cause," the *vera causa*, which would become the hallmark of so many claims of seventeenth-century science. Galileo was not searching for the cause that Aristotle had assigned, he was searching for the "true cause" of falling motion. Returning to the same terminology later, he makes the even bolder assertion:

> We have no hesitation in asserting that this is the true, essential, and foremost cause that explains why natural motion is slower at the beginning. Those who examine it properly and thoroughly will no doubt accept it and embrace it as completely true. [87]

That this is no isolated instance, finally, becomes clear when we peruse Chapter 23, where Galileo is inquiring into "why objects projected by the same force move farther on a straight line the less acute are the angles they make with the plane of the horizon," [88] and which he illustrates with the example of iron balls shot from cannons to batter down fortifications. Here his announced intention is to "find the true cause of this effect, whatever others may say." [89] Thus there can be no doubt that in this early *De motu* Galileo was convinced that local motion could be analyzed in terms of its proper causes and was in fact searching to discover what these causes might be.

b. Galileo's Methodology

In the *De motu*, in his later work at Padua, and also in his writing up to and including the *Dialogues Concerning the Two Chief World Systems*, Galileo's methodology explicitly employed a resolutive and compositive method that is distinctively Aristotelian, [90] although the Florentine physicist made little effort at this stage of his life to acknowledge any debt to the Greek philosopher. In this respect his method was not appreciably different from those we have seen in the Paduan school to the end of the sixteenth century. What is more characteristic of Galileo, however, is his emphasis on mathematical reasoning in the Platonic and Archimedean mode, his insistence that the book of nature "is written in the mathematical language." [91] Perhaps because he had been a professor of mathematics, he regarded this science as more important than logic in scientific investigation.[92] He was thus convinced that mathematics supplies an apt model for all scientific work; more than this, he felt that it would also serve as an instrument of discovery for uncovering the true causes of natural phenomena. Although sympathetic toward Gilbert, moreover, he in fact criticized the great British experimentalist for failing to make sufficient use of mathematical reasoning in his search for causes:

> He [Gilbert] seems to me worthy of great acclaim also for the many new and sound observations which he made . . . [But] what I might have wished for in Gilbert would be a little more of the mathematician, and especially a thorough grounding in geometry, a discipline which would have rendered him less rash about accepting as rigorous proofs those reasons which he puts forward as *verae causae* for the correct conclusions he himself had observed. His reasons, candidly speaking, are not rigorous, and lack that force which must unquestionably be present in those adduced as necessary and eternal scientific conclusions.[93]

Granted Galileo's faith in mathematical insight, his methodology was enhanced also by an interest in experimental method and by an acquaintance with the developing craft tradition that was a forerunner of modern technology. Whether Galileo himself used experimentation as an instrument of discovery, however, particularly in his early years, is much disputed among scholars.[94] He seems not to have been as trustful of sense knowledge as the British empiricists who would follow him within a century or so, creating the impression that putting scientific questions to sensory test will be powerless, in most instances, to reveal a correct causal explanation.[95] It would appear that experimentation performed largely a negative function in Galileo's early work, even though he did make a distinction between ordinary experience and experiment in the more scientific sense. One of his criticisms of Aristotle, which we have already mentioned, is that Aristotle relied too much on experience, and Galileo saw this as powerless to reveal the causes of natural phenomena.

In the *De motu* he is explicit that "what we seek are the causes of effects, and these causes are not given to us by experience . . ."[96]

Even in his later writings, as in the *Dialogi*, what appear to be experiments in the modern mode are frequently only imaginary or thought experiments.[97] A fascinating case is the discussion, in the *Dialogi*, of the weight dropped from the mast of a ship, when the ship is first at rest and then in motion, to see if in the latter case the weight lands an appreciable distance behind the foot of the mast. Here both the Aristotelian Simplicio and his adversary Salviati adduce experimental evidence in favor of their preconceived conclusions, while admitting, on cross-examination, that neither has actually performed any experiments, and, in Salviati's case, that he knew the result with such assurance beforehand that experiments were not necessary.[98] Later on, in the *Discorsi*, Galileo would himself make the more explicit claim that "the knowledge of a single fact acquired through a discovery of its causes prepares the mind to understand and ascertain other facts without need of recourse to experiment . . ."[99]

Whether Galileo's views on methodology underwent substantial modification in his declining years is a further topic of debate among scholars. There are some who see in the passage with which we opened this chapter, and even in the discussion of the meaning of the term "gravity" in the *Dialogi*,[100] a tendency toward positivism, or at least toward agnosticism with respect to causal explanations.[101] Stillman Drake argues, in this connection, that it was precisely for his "mature rejection of the quest for causes in physics" that Galileo came under criticism from the great contemporary philosopher and systematizer, René Descartes.[102]

Other evidence is available, however, to indicate that Galileo's agnosticism, if it be such, was more a methodological device to guard against premature conjectures based on insufficient evidence, and that, up to his last years, he was still intent on discerning the causes and rational explanations of natural phenomena, much as Sir Isàac Newton would later prove to be. The structure of his final work, the *Discorsi*, lends support to this interpretation. The discussions of the first two days, concerned as they are with the mechanical resistance of bodies, or what would later be referred to as the strength of materials, is actually a quest for the hidden cause of the cohesion to be found among the parts of solid bodies.[103] Then, in the third and fourth days, Galileo presents his new science of motion after the fashion of a classical, deductive exposition, where the basic principles have already been ascertained and the new science is to be concerned with deducing properties or effects from these principles. Unlike the dialogues of the first two days, the third and the fourth days are devoted to the reading of a Latin treatise, *De motu*, which is commented on by the discussants. Obviously the work of Galileo himself, this revised *De motu* follows a mathematical form of exposition, and the only place where experience or experimentation enters is when the discussants question the correspondence between the results demonstrated in the treatise and the facts of experience as they know them. Predictably, it is the Aristotelian Simplicio who is the most concerned with experiential reference and with experimental verification. Sagredo and Salviati, who follow Galileo's own line of thought quite sympathetically, have a more mathematical and a less empirical orientation. Thus Sagredo affirms that to have a mathematical understanding of the causes of an event "far outweighs the mere information obtained from the testimony of others or even by repeated experiment." [104] It is here that Salviati adds the comment to which we have already made reference:

> What you say is very true. The knowledge of a single fact [effect] acquired through a discovery of its causes prepares the mind to understand and ascertain further facts [effects] without need of recourse to experiment . . .[105]

When these statements are taken in conjunction with Galileo's general philosophy, wherein he subscribed to an atomic theory of matter,[106] and used this to argue for the subjectivity of secondary qualities, they lead one to believe that Galileo, no less than Kepler or Gilbert, was intent on providing causal explanations for natural phenomena. His attempts at understanding how the inner constituents of bodies, whether these be conceived as *minima* or atoms, could be in motion and so might act on the human body to explain man's sensations, inevitably led him to speculate about mechanical causes in ways that would influence Descartes and other philosophers we shall study in the next volume. The discovery of such causal mechanisms in fact constituted Galileo's ideal of science, and this whether the causes they involved could be discovered by mathematical insight or would yield themselves also to experimental investigation.[107] If Galileo differed from Aristotle in this matter, it was over precisely what would constitute such causes, and not over the search itself. It is in such a context that one can have sympathy with Geymonat's statement to the effect that Galileo wished "to recognize the existence of a profound linkage between [his] new science and the best parts of Aristotelian thought." [108] Geymonat goes on:

> Indeed, when in 1640 he again discussed the relations between his own methodology and that of Aristotle, Galileo admitted the existence of a real link between them. He even went so far as to assert that he, rather than his adversaries, was the true heir of Aristotle.[109]

It may be difficult to reconcile assertions such as this with the Platonism and the positivism that have been variously ascribed to Galileo, but they are not inconsistent with the polemics against the Aristotelians in Galileo's previous writings. The failure of the sixteenth-century followers of Aristotle was not that they had searched for causes, but rather that they had given up the search too easily and halted their inquiry short of the "true causes" now about to be unveiled through the "new science" Galileo himself had finally come to propose.

4. William Harvey

Whether Harvey knew Galileo personally is not known, but the two were contemporaries at the University of Padua, where Harvey was awarded the doctorate in 1602 as "most ef-

ficiently qualified both in arts and medicine . . ." [110] Unlike
Galileo, however, Harvey was consistently a great admirer of
Aristotle, consciously employing and defending the methodology of the *Posterior Analytics* as an instrument in his scientific researches. This is not to say that Harvey was uncritical
of the Aristotelians and the Galenists of his day; he too could
inveigh against those who wasted their time quoting the
opinions of the ancients and were unwilling to learn from the
book of nature itself. Harvey was no more content to rest
with the conclusions of the early Greeks than was Galileo,
but he saw his own conclusions, novel though they were to
the readers of his day, as simply the result of the patient application of Aristotelian methodology to the study of natural
processes. Both of his classical treatises, *An Anatomical Disquisition on the Motion of the Heart and Blood in Animals*
and *Anatomical Exercises on the Generation of Animals*, were
in fact proposed as exercises in the application of precisely
this methodology.[111]

a. Aristotelian Demonstration

The first of these works was composed when Harvey was
about fifty years old, whereas the second was published
twenty-three years later and shows how unvarying he was,
even in his last years, with respect to the methodological positions he had adopted. The second work, in fact, begins with
an essay on scientific method that is solidly Aristotelian and
invokes the classical texts of the *Physics* and *Posterior Analytics* dealing with induction and its role in the demonstrative
process. Harvey is there insistent that the scientist must base
his work on "his own experience, i.e., from repeated memory,
frequent perception by sense, and diligent observation, [he
must] know that a thing is so in fact." [112] It is the failure to
do this that has led many to "only imagine or believe" without having true science themselves.[113] This is the context in
which Harvey disparages the method then practiced of the
schools, in the following terms:

> The method of investigating truth commonly pursued at
> this time, therefore, is to be held as erroneous and almost
> foolish, in which so many inquire what others have said,
> and omit to ask whether the things themselves be ac-

tually so or not; and single universal conclusions being deduced from several premises, and analogies being thence shaped out, we have frequently mere verisimilitudes handed down to us instead of positive truths.[114]

As opposed to this, Harvey advises his readers to have recourse to the methods of observation and verification practiced by Aristotle, and even has reference to a text we have already pointed out where Aristotle advocates a procedure overlooked by many later Aristotelians.[115] Harvey's own advice is yet more insistent:

> I, therefore, whisper in your ear, friendly reader, and recommend you to weigh carefully in the balance of exact experience all that I shall deliver in these *Exercises on the Generation of Animals;* I would not that you gave credit to ought they contain save insofar as you find it confirmed and borne out by the unquestionable testimony of your own senses.[116]

He then concludes his introductory disquisition on method by showing how "inquiry must begin from the causes, especially the material and efficient ones," [117] and lays out his program for causal analysis based on careful observation and experimentation. Mindful of his debt to his predecessors, however, he tells the reader that "I would have you know that I tread in the footsteps of those who have already thrown a light upon this subject, . . . and foremost of all among the ancients I follow Aristotle; among the moderns, Fabricius of Aquapendente; the former as my leader, the latter as my informant of the way." [118]

The same tone and procedure also characterizes Harvey's much briefer classic *On the Motion of the Heart and Blood.* This work is more demonstrative than descriptive in intent, the greater part of it being directed toward establishing "the conclusion of the demonstration of circulation [of the blood]." [119] Although not published until 1628, the essential elements of the demonstration were already contained in notes composed by Harvey for a series of lectures given at the Royal College of Physicians in London in 1616.[120] Both in these lectures and in the written work, Harvey argued against a deeply entrenched and commonly entertained view that blood was produced in a central organ within the body and then distributed to the periphery, being totally consumed in the process. Working before the discovery of the microscope, and thus unable to trace the complete course of the blood's circulation, he was nonetheless able to demonstrate from the valve action and from the quantity of blood contained in the body that continuous motion in a circle is the only way of accounting for the blood's flow. From this it was a simple matter for him to conclude to the causes of this circulation and thus to connect the flow of blood with the pumping action of the heart.

Harvey's classic is so tightly written that it is easy to discern its logic from perusal of the titles of the seventeen chapters of which it is composed. The dedicatory letters, the introduction, and Chapter 1 are all directed toward establishing the need for the work, considering particularly the many erroneous opinions being held and the difficulty of setting these aright. The next main division, made up of Chapters 2 through 7, is an extended observational analysis of the motion of the cardiovascular system, based on the dissection of living animals, with attention being given successively to the motion of the containing parts, i.e., the arteries and the heart, as well as the motion of the contained part, i.e., the blood, with particular reference to the ventricles of the heart and the passage through the lung. With these facts established, Harvey builds his demonstration of the circular motion of the contained part in Chapters 8 through 14, basing it on the thesis that the abundance of blood passing through the heart out of the veins into the arteries can only be accounted for by the blood's circular motion. Many of the observations he uses to establish his point had been known to his predecessors, but none was able to put them in the precise logical order necessary to explain all the facts. Harvey was himself conscious of the great mass of evidence he had accumulated, and of the difficulty of arranging this in a way that would be acceptable and understandable to his contemporaries, in view of the "novel and unheard of character" of the explanation to which he had come.[121] But he states his case nonetheless, and in so doing provides a rare insight into the process of scientific discovery:

Still the die is cast, and my trust is in my love of truth, and the candour that inheres in cultivated minds. And sooth to say, when I surveyed my mass of evidence, whether derived from vivisections, and my various reflections on them, or from the ventricles of the heart and the vessels that enter into and issue from them, the symmetry and size of these conduits — for nature doing nothing in vain, would never have given them so large a relative size without a purpose — or from the arrangement and intimate structure of the valves in particular, and of the other parts of the heart in general, and with many things besides, I frequently and seriously bethought me, and long revolved in my mind, what might be the quantity of blood which was transmitted, in how short a time its passage might be effected, and the like; and not finding it possible that this could be supplied by the juices of the ingested aliment without the veins on the one hand becoming drained, and the arteries on the other getting ruptured through the excessive charge of blood, unless the blood should somehow find its way from the arteries into the veins, and so return to the right side of the heart; I began to think whether there might not be a *motion, as it were, in a circle.* Now this I afterwards found to be true . . .[122]

Once Harvey had hit upon this insight he saw immediately why, in dead animals, "we usually find so large a quantity of blood in the veins, so little in the arteries . . . much in the right ventricle, little in the left . . ." [123] "The true cause of the difference is this," he explains, that there is no passage through the arteries except through the lungs and the heart, and so when an animal ceases to breathe and the lungs to move, the source of supply to the veins is cut off, although the heart, "surviving them and continuing to pulsate for a time," causes blood to accumulate in the veins.[124] Seeing that "now the cause is manifest," [125] Harvey then repeats the experiments performed by his teacher Fabricius with ligatures applied to the arm of a living man, and shows that these too are explainable in the light of the circulation of the blood.[126] Taking all of this into account he is finally able to conclude, with absolute certitude, that his is the one and only correct explanation of the motion of the heart and blood in animals.

He sums up his argument in Chapter 14, which is itself a brief recapitulation of the demonstration and thus is worth citing in its entirety:

> And now I may be allowed to give in brief my view of the circulation of the blood, and to propose it for general adoption. Since all things, both argument and ocular demonstration, show that the blood passes through the lungs and heart by the action of the auricles and ventricles, and is sent for distribution to all parts of the body, where it makes its way into the veins and pores of the flesh, and then flows by the veins from the circumference on every side of the center, from the lesser to the greater veins, and is by them finally discharged into the vena cava and right auricle of the heart, and this in such a quantity or in such a flux and reflux thither by the arteries, hither by the veins, as cannot possibly be supplied by the ingesta, and is much greater than can be required for mere purposes of nutrition; *it is absolutely necessary to conclude* that the blood in the animal body is impelled in a circle, and is in a state of ceaseless motion; that this is the act or function which the heart performs by means of its pulse; and that *it is the sole and only end* of the motion and contraction of the heart.[127]

The reasoning is lucid and precise; Harvey knows he has a demonstration based on causal analysis, and thus that the fact of the blood's circulation could not be other than it is, and so he states his conclusion with uncompromising accuracy and conviction.

Having thus concluded apodictically, Harvey devotes the remaining three chapters to various a posteriori proofs [128] and to confirmatory arguments. His final chapter serves also as a summary and synthetic exposition of the definition of the heart, touching on all four of its causes, viz.: (1) its formal cause, the anatomical structure described in terms of its function; (2) its material cause, the muscular and other tissue sustaining this structure and operation; (3) its final cause, the circulation of the blood; and (4) the efficient cause of the circulation, the contraction whereby the heart fulfills its function. Having explained all this in detail Harvey concludes his treatise simply with the words, "it would be difficult to explain in any other way for what cause all is constructed and arranged as we have seen it to be." [129]

b. Causal Explanations

Although a classical demonstration in the Aristotelian mode, Harvey's treatise *On the Motion of the Heart and Blood* was misunderstood in its day, as it continues to be in our own.[130] René Descartes, who knew of Harvey's work and was generally sympathetic to it, failed to grasp the demonstrative force of the argument. Francis Bacon, although served by Harvey as his personal physician, gave no indication that he even recognized Harvey's contribution to biological science; in the words of a leading Bacon scholar, "the probability is that . . . he regarded [Harvey's] theory as hardly worthy of serious discussion." [131] The reception among lesser known thinkers was no better, with the result that Harvey's great discovery was greeted with almost universal skepticism by his contemporaries. One such contemporary, a Frenchman named Jean Riolan, criticized Harvey's exposition in his *Encheiridium anatomicum et pathologicum*, published at Paris in 1648, and succeeded in eliciting two replies from Harvey. These deserve comment for the further light they shed on Harvey's methodology with regard to two important points, viz., (1) the role of vital spirits in causal explanations of living organisms, and (2) the completeness of explanations that can be expected in terms of efficient and final causes.

The *Second Disquisition to Riolan*, in particular, adduces further experiments that lay bare "the causes and reasons" [132] of the blood's circulation. In his first reply Harvey had made clear "the principal use and end of the circulation," namely, to enliven all of the parts of the body, which, "to use the language of physiologists," are sustained and actuated "by the inflowing heat and vital spirits." [133] Apparently Riolan was impressed by Harvey's reference to "vital spirits" and sought in these a possible explanation of the blood's movement. In his *Second Disquisition* Harvey takes up the question of such spirits, acknowledging that "it is still a question what they are," and that upon this point "there are so many and such conflicting opinions, that it is not wonderful that the spirits, whose nature is thus left so wholly ambiguous, should serve as the common subterfuge of ignorance." [134] He goes on:

> Persons of limited information, when they are at a loss to assign a cause for anything, very commonly reply that it is done by spirits; and so they bring the spirits into play upon all occasions; even as indifferent poets are always thrusting the gods upon the stage as a means of unravelling the plot, and bringing about the catastrophe.[135]

He then surveys various opinions on the different kinds of spirits spoken of in medical schools, noting that "there is nothing more uncertain and questionable . . . than the doctrine of spirits that is proposed to us," [136] and concluding with the remark that "the spirits which flow by the veins or the arteries are not distinct from the blood, any more than the flame of the lamp is distinct from the inflammable vapour that is on fire . . ." [137]

Notwithstanding Harvey's rather thorough causal analysis, particularly as seen in his concluding chapter with its definition of the heart and its functions, he was quite prepared to admit that he did not have complete explanations, especially in the matter of the final and efficient causes of the circulation. Riolan seems to have touched on this problem, and Harvey sees in this an opportunity to reply to his and other criticisms. He writes:

> To those who repudiate the circulation because they neither see the efficient nor final cause of it, and who exclaim *cui bono*, I have yet to reply, having hitherto taken no note of the ground of objection which they take up. And first I own I am of opinion that our first duty is to inquire whether the thing be or not [*an sit*] before asking wherefore it is [*propter quid*], for from the facts and circumstances which meet us in the circulation admitted, established, the ends and objects of its institution are especially to be sought.[138]

Harvey's answer to such critics is simple: he would merely have them turn their attention first to establishing the fact before seeking the reasoned fact, even though he had already established both fact and reasoned fact with as much clarity as was possible in his day. As to ascertaining the fact, Harvey's further advice to his critics is to avoid erroneous opinions, no matter how venerable they may be, and to employ only the manifest data of the senses. For "the facts manifest by the senses wait upon no opinions, and . . . the works of nature bow to no antiquity; for indeed there is nothing either more ancient or of higher authority than nature." [139] Here his emphasis is again on the importance of accurate sense observation to establish the circulation of the blood. Harvey realizes, of course, that sense and reason must cooperate in this enterprise, and that there are some instances in which reasoning must make up for the deficiencies of sense experience. In this context he discusses the case, proposed by Aristotle in the *Posterior Analytics,* of the respective roles of sense and intellect in understanding the lunar eclipse,[140] but warns that it is not necessary to use the methods of astronomers when dealing with medical matters that are readily available to sense investigation. So he cautions:

> And here the example of astronomy is by no means to be followed, in which from mere appearances or phenomena that which is in fact, and the reason wherefore it is so, are investigated. But as he who inquires into the cause of an eclipse must be placed beyond the moon if he would ascertain it by sense, and not by reason, still, in reference to things sensible, things that come under the cognizance of the senses, no more certain demonstration or means of gaining faith can be adduced than examination of the senses, than ocular inspection.[141]

Harvey thereupon repeats the demonstrative argument contained in his work *On the Motion of the Heart and Blood,* the premises of which he had established by observation and experimentation with the utmost care. He concludes:

> Now this, my conclusion, is true and necessary, if my premises be true; but that these are either true or false, our senses must inform us, not our reason — ocular inspection, not any process of the mind.[142]

It should be stressed that Harvey's empiricism is not meant to exclude a further search for causes, but merely to determine the facts on which such a search can intelligently be based. For example, having established the fact of the blood's circulation, he was quite certain that the efficient cause of this circulation is the contraction of the muscles of the heart. Thus, for him, the efficient cause of the circulation is the heart's pumping action.[143] What the further efficient cause of the pumping might be, however, was not clear to him, although he discusses this type of question in the last part of his *Second Disquisition to Riolan,* immediately after having stressed the importance of visual inspection for establishing the basic fact he is attempting to explain. He thereupon notes, "without pretending to demonstrate it," [144] his own explanation of the cause of the heart's motion, disagreeing with most of the opinions expressed in his day, including that of Descartes, and concluding "that the rising and falling of the blood does not depend upon vapours and exhalations, or spirits, or anything arising in a vaporous or aerial shape, nor upon any external agency, but upon an internal principle under the control of nature." [145] The "internal principle" to which he refers would be recognized by his contemporaries as the soul, or *anima,* the formal principle that satisfies Aristotle's definition of nature as "the primary source of movement and rest" in animated things.[146]

As we shall see, Harvey's search for the causes of the circulation of the blood thus terminates in much the same way as does Newton's search for the causes of gravitational motion and of the colors of the spectrum. Both employ induction and demonstrative reasoning to establish what they regard as a certain conclusion, for which they propose a proper or proximate causal explanation. When searching deeper for a more ultimate cause that might in turn explain the proper and proximate cause, however, they make no extravagant claims. And yet they will not allow one who is vexed at the inability to supply such a deeper, underlying cause, to deny the demonstrative character of the argument based on the proper and proximate cause they have discovered. Thus, in effect, they open up the possibility of continued advance in scientific knowledge at the level of proximate causes, while not requiring their followers to embrace a total systematic explanation that would pretend to say the last word with regard to the ultimate causes of all phenomena.

5. Isaac Newton

Sir Isaac Newton is a key figure in the foundation of modern physical science, for it was he who systematized in a conceptual way the discoveries of Galileo and Kepler and showed how the principles of mechanics could be extended to a discussion of most topics of interest to natural philosophers, from the structure of matter all the way to the structure of the universe.[147] Newton, like Harvey, was also much interested in methodology, and attempted time and time again to clarify and defend the methods by which he got his results. Like Harvey again, Newton produced two classics, *The Mathematical Principles of Natural Philosophy,* wherein are laid the foundations of classical mechanics, and the *Opticks,* a somewhat less systematic work but containing nonetheless the principles on which much of the later study of light and color was to be based. Both employ an inductivist approach — and this too a note of similarity with Harvey — which Newton felt enabled him to discern the "true cause" of certain gravitational and optical phenomena, without pretending in so doing to give ultimate answers regarding the nature of gravity or of light and color.

The sources of Newton's methodological convictions have not been thoroughly studied to date, but there can be little doubt that his method of analysis and synthesis is closely related to the processes of resolution and composition that derive from the *Posterior Analytics*. By the time he went to study at Cambridge, in 1661, a number of modifications had already been introduced into the curriculum there, so that his was not quite the same type of education as Galileo and Harvey had received at Pisa and Padua respectively. Still, much of the late medieval curriculum remained, and Newton began his undergraduate study with the traditional courses in logic, rhetoric, and ethics in the Aristotelian mode.[148] His early student notebooks show also an acquaintance with Aristotelian natural philosophy, particularly as revealed in his discussion of violent (or projectile) motion, although not to the extent of Galileo's *Juvenilia* and *De motu*. Of Newton's notations in his *Wastebook* Herivel observes:

This considerable discussion of violent motion is memorable as being the earliest extant extended piece of writing by Newton on a dynamical subject. It is also of great interest for the evidence it provides for the medieval background to Newton's thought. The tone of the whole passage is unmistakably scholastic, being cast almost in the form of a medieval disputation.[149]

Much of this paper, predictably enough, is concerned with discovering the cause of the continuation of a projectile's motion, of which Newton recognizes three distinct possibilities: (1) the air, (2) a "force impresst," and (3) the "natural gravity" of the body. Newton rejects the first two possible causes in the first section of the paper, and devotes the remainder to "a defence of the third, and, in [his] view, correct cause." [150] It is to be noted, of course, that Newton did not persist in peripatetic discussions of this type, for by 1664 he had begun a new set of notes based on the writings of Galileo and Descartes, among others, along the lines of the new mechanical philosophy.

While these *Quaestiones quaedam philosophicae,* the title Newton appended to the new notes, are a fascinating source of study for the origins of his dynamical principles and concepts, they are also revealing for their preliminary analysis of color phenomena, for it is here that his distinctive methodological views first began to take shape.[151] The problem of colors had early become a test case for advocates of the mechanical philosophy; as we shall see in the next volume, Descartes had already attempted to explain colors as modifications of light associated with a rotary motion of the particles of which he thought light was composed. Newton saw defects in this explanation, and by 1666 he had hit upon a new conception of color formation which he succeeded in subjecting to experimental test. In 1670 he was appointed to succeed Isaac Barrow in the chair of mathematics at Cambridge, and devoted his first series of lectures, the *Lectiones opticae,* to the problems of light and color. By 1672 Newton in effect had written the entire *Opticks,* but he delayed publication of this until 1704, undoubtedly because of his extreme sensitivity with regard to criticisms that attended his first effort to explain his views of light and color.[152]

a. The *Experimentum crucis*

This initial effort, actually Newton's first published account of a discovery, was printed in the *Philosophical Transactions* of 1671/72; it was in the form of a letter to the Secretary of the Royal Society, Henry Oldenburg, who read it to the Society on February 8, 1671/72.[153] In the letter, wherein Newton describes his discoveries and experiments as taking place in a simple chronological sequence during the year 1666, he appears quite convinced that he has transcended the hypothetical type of explanation offered by Descartes and others and has succeeded in giving a true causal explanation of the formation of colors. In a section of his letter not printed in the *Transactions,* in fact, Newton writes:

> A naturalist would scarce expect to see the science of those [colors] become mathematical, and yet I dare affirm that there is as much certainty in it as in any other part of optics. For what I shall tell concerning them is not an hypothesis but most rigid consequents, not conjectured by barely inferring 'tis thus because not otherwise, or because it satisfies all phenomena (the philosophers' universal topic), but evinced by the mediation of experiments concluding directly and without any suspicion of doubt.[154]

From this passage and from other indications,[155] historians of science question whether Newton's account is as chronological as he presents it, and propose instead that it is more a methodological reconstruction that presents "an argument designed to establish the final doctrine beyond all possible objections." [156] There can be no doubt, however, that Newton here sets himself irrevocably against a mere hypothetical methodology, and in its place proposes a strict demonstration based on observation and experimentation.

The key to understanding Newton's demonstration, which also explains how in the light of his discovery optical science had "become mathematical," in his observation of a peculiarity in the shape of the spectrum that results when a beam of light passes through a prism and is projected on an opaque surface. What first caught Newton's attention, if we are to go by his account, was that the circular beam, which he thought should project an orbicular image, assumed an elongated or "oblong form" whose length was about five times greater than its breadth — "a disproportion so extravagant," Newton writes, "that it excited me to a more than ordinary curiosity of examining from whence it might proceed." [157] Acquainted as he was with the various mechanistic theories of light current in his day, Newton apparently first tried to account for the elongation of the image in terms of explanations such theories might afford. Thus it occurred to him that perhaps the shape of the spectrum was explicable in terms of the thickness of the prism or the size of the aperture through which the circular beam of sunlight was initially admitted to it. Another possibility was that the disproportionate length of the image could be caused by some unevenness or irregularity in the glass of the prism. Yet a third was that light might be composed of small particles that rotate, along the lines of Descartes' conception, and that spinning motions of varying degrees might be imparted to the light particles by their passage through different parts of the prism, causing them to travel in lines of varying curvatures and thus distend the resulting image.

Each of these hypotheses Newton considers in turn and falsifies by a series of remarkable experiments.[158] Having done this, if we are to give credence to his autobiographical account, he begins to speculate as to what the "true cause" of the elongation of the image might be. This leads him to what he calls his *experimentum crucis*, a crucial experiment performed with two prisms set up in such a way that he can control a particular ray of light passing through the second prism and see precisely what the relationship is between the color of the ray and the angle at which it is refracted. He performs the experiment and notes that the red rays at one end of the spectrum are refracted very little, whereas the violet rays at the opposite end "suffer a refraction considerably greater . . ." [159] Noting this difference, Newton sees in it the proper cause that explains why the image is elongated, and concludes immediately:

> And so the true cause of the length of that image was detected to be no other, than that light consists of rays dif-

ferently refrangible, which, without any respect to a difference in their incidence, were, according to their degrees of refrangibility, transmitted towards divers parts of the wall.[160]

The conclusion of the *experimentum crucis*, therefore, is that sunlight or white light is composed of different rays that are refractible in varying degrees, and that the resolution of the white light into its various components by the prism is the only way of accounting for the elongated shape that is projected on the wall.

A methodological analysis of the brief passage in which Newton describes this *experimentum crucis* shows that he implicitly utilized a twofold process of resolution and composition, or, as he prefers to call it, analysis and synthesis.[161] The first resolves the elongation of the image into the various components, or colored rays, of which he finds white light to be composed, and thus reduces the phenomenon of the image's elongation to its proper cause. The composition then takes this cause, viz., the fact that white light is composed of rays with different degrees of refractibility, and explains how the passage through the prism spreads out these components and thus produces the effect that excited Newton's "more than ordinary curiosity" in the first place.[162]

The quantitative aspect of the image's elongation, for which Newton sought the "true cause," leads him on further reflection to a qualitative conclusion respecting radiant color and its relationship to white light. Thus he concludes from this experiment:

> Colours are not qualifications of light, derived from refractions or reflections of natural bodies (as 'tis generally believed) but original and connate properties, which in divers rays are divers. Some rays are disposed to exhibit a red color and no other; some a yellow, and no other; some a green, and no other; and so of the rest. Nor are there only rays proper and particular to the more eminent colours, but even to all their intermediate gradations.[163]

Moreover, not only are colors proper to the rays of which white light is composed but the angle of refraction that the ray experiences when passing through the prism is also a property, immutably connected with the ray's color. Newton goes on:

> The species of colour, and degree of refrangibility proper to any particular sort of rays, is not mutable by refraction, nor by reflection from natural bodies, nor by any other causes that I could yet observe. When any sort of rays have been well parted from those of other kinds, it hath afterwards obstinately retained its colour, notwithstanding my utmost endeavours to change it. I have refracted it with prisms, and reflected it with bodies, which in daylight were of other colours; I have intercepted it with a colored film of air, interceding two compressed plates of glass, transmitted it through mediums, and through mediums irradiated with other sorts of rays, and diversely terminated it; and yet could never produce any new colour out of it.[164]

Thus, if we are to take Newton's account literally, he tried every possibility at his command to change the color of a ray associated with a particular degree of refraction, but was never able to effect such a change. As far as he could tell, therefore, he had discovered a true property of the rays that go to make up white light.

It is most interesting to study the various reactions evoked by the publication of Newton's first paper in the *Philosophical Transactions*. The general tenor of the responses is one of nonacceptance, and this, as Rosenfeld has surmised, because of a failure on the part of Newton's adversaries to comprehend the method he had used to establish his results.[165] Criticisms were voiced by such eminent scientists as Robert Hooke and Christian Huygens, and by French and English Jesuits on the Continent, who were under the influence of their confrère Grimaldi, himself a famous optician. All subscribed to a hypothetical system of explanation wherein they accounted for the various properties of light and color through one or other mechanical hypothesis. Newton, on the other hand, deliberately avoided such hypotheses, as he was later to avoid them in his explanations of gravity. On this account he was suspected of peripatetic tendencies, since he seemed to prefer qualities and "original properties" to the

Cartesian type of explanation that had become popular throughout all of Europe.

As it turns out, Newton's critics were seeking more ultimate explanations than he was, and so he consistently answered them by saying that he was not committing himself on the nature of either light or color, but merely wished to demonstrate properties that could be experimentally verified. Thus, when Hooke charged Newton with holding that light is a material substance, Newton replied that this was not his intention, but that he intended "to speak of *light* in *general* terms, considering it abstractly, as something or other propagated every way in straight lines from luminous bodies, without determining, what that thing is; whether a confused mixture of difform qualities, or modes of bodies, or of bodies themselves, or of any virtues, powers, or beings whatsoever. And for the same reason I chose to speak of *colours* according to the information of our senses, as if they were qualities of light, *without* us." [166] Similarly Huygens reproached Newton for not having taken account of "an hypothesis of motion . . . , and till he hath found this hypothesis, he hath not taught us, what it is wherein consists the nature and difference of colours, but only this accident (which certainly is very considerable) of their different refrangibility." To this Newton again replied: "I never intended to show wherein consists the nature and difference of colours, but only to show that *de facto* they are original and immutable qualities of the rays which exhibit them." [167] He went on to explain that the most he would conclude about colors is that they themselves are basic and irreducible qualities, and that he would not attempt to explain their varieties in any deeper way, merely explaining them through an effect or property that accompanies such qualities whenever they appear.

More revealing for the present study are Newton's replies to the Jesuits who criticized his paper, possibly because he felt he could presume in them a better knowledge of demonstrative methodology. For example, the French Jesuit Ignace Pardies wrote to the Royal Society about "Mr. Newton's very ingenious hypothesis of light and colours," [168] treating all of Newton's exposition as merely hypothetical. To this Newton replied immediately, disavowing any hypothetical character to his explanation:

> I do not take it amiss that the Rev. Father calls my theory an hypothesis, inasmuch as he was not acquainted with it. But my design was quite different, for it seems to contain only certain properties of light, which, now discovered, I think easy to be proved, and which if I had not considered them as true, I would rather have them rejected as vain and empty speculation, than acknowledged even as an hypothesis.[169]

This observation, together with the detailed way in which Newton met Pardies's arguments and objections, elicited from the French Jesuit the gracious response: ". . . as to my calling the author's theory an hypothesis, that was done without any design, having only used that word as first occurring to me; and therefore request it may not be thought as done out of any disrespect."[170]

Still Pardies had other difficulties, and his formulation of these brought from Newton a yet fuller explanation of the *experimentum crucis*. Pardies, as Newton analyzed his objections, was still under the impression that the experiment could be explained without recourse to Newton's "true cause" but was reconcilable with one or other mechanical hypothesis, such as those of Grimaldi, Hooke, or Descartes. So Newton answers again that no hypothesis of any kind was involved in the experiment:

> In answer to this, it is to be observed that the doctrine which I explained concerning refraction and colours consists only in certain properties of light, without regarding any hypotheses by which those properties might be explained. For the best and safest method of philosophizing seems to be, first to inquire diligently into the properties of things, and establishing those properties by experiments, and then to proceed more slowly to hypotheses for the explanation of them. For hypotheses should be subservient only in explaining the properties of things, but not assumed in determining them; unless sofar as they may furnish experiments. For if the possibility of hypotheses is to be the test of the truth and reality of

things, I see not how certainty can be obtained in any science; since numerous hypotheses may be devised, which shall seem to overcome new difficulties. Hence it has been here thought necessary to lay aside all hypotheses, as foreign to the purpose, that the force of the objection should be abstractly considered, and receive a more full and general answer.[171]

Pardies seems to have been satisfied with this explanation, and particularly with the more complete diagrams of his experiment with which Newton had supplied him, and thus he concluded the correspondence with the following note to the Royal Society:

I am quite satisfied with Mr. Newton's new answer to me. The last scruple which I had, about the *experimentum crucis,* is fully removed. And I now clearly perceive by his figure what I did not before understand. When the experiment was performed after his manner, every thing succeeded, and I have nothing further to desire.[172]

Of similar interest is Newton's interchange with two British Jesuits who were then teaching at the English college of Liège, Francis Line and Anthony Lucas.[173] Lucas, in particular, seems to have been a meticulous experimenter, and wrote Newton about the difficulties he experienced with the *experimentum crucis,* suggesting in the process that Newton perform yet more experiments. The following lengthy excerpt from Newton's answer to Lucas is of particular value for its insistence upon the demonstrative character of the experiment, and the way in which its proper performance will lead the English Jesuit to a grasp of the truth. Newton's communication reads as follows:

Concerning Mr. Lucas's other experiments, I am much obliged to him that he would take these things so far into consideration, and be at so much pains for examining them; and I thank him so much the more, because he is the first that has sent me an experimental examination of them. By this I may presume he really desires to know what truth there is in these matters. But yet it will conduce to his more speedy and full satisfaction if he a little change the method which he has propounded, and instead of a multitude of things try only the *experimentum crucis*. For it is not number of experiments, but weight to be regarded; and where one will do, what need many?

Had I thought more requisite, I could have added more: For before I wrote my first letter to you about colours, I had taken much pains in trying experiments about them, and written a tractate on that subject, wherein I had set down at large the principal of the experiments I had tried; amongst which there happened to be the principal of those experiments which Mr. Lucas has now sent me. And as for the experiments set down in my first letter to you, they were only such as I thought convenient to select out of that tractate.

But suppose those had been my whole store, yet Mr. Lucas should not have grounded his discourse upon a supposition of my want of experiments, till he had examined those few. For if any of those be demonstrative, they will need no assistance, nor leave room for further disputing about what they demonstrate.

The main thing he goes about to examine is, *the different refrangibility* of light. And this I demonstrated by the *experimentum crucis*. Now if this demonstration be good, there needs no further examination of the thing; if not good, the fault of it is to be shown: for the only way to examine a demonstrated proposition is to examine the demonstration. Let that experiment therefore be examined in the first place, and that which it proves be acknowledged, and then if Mr. Lucas wants my assistance to unfold the difficulties which he fancies to be in the experiments he has propounded, he shall freely have it; for then I suppose a few words may make them plain to him: whereas, should I be drawn from demonstrative experiment to begin with those, it might create us both the trouble of a long dispute, and by the multitude of words, cloud rather than clear up the truth. For if it has already cost us so much trouble to agree upon the matter of fact in the first and plainest experiment, and yet we are not fully agreed; what an endless trouble might it create us, if we should give ourselves up to dispute upon every argument that occurs, and what would become of truth in such a tedious dispute? The way therefore that I propound, being the shortest and clearest (not to say, the only proper way), I question not but Mr. Lucas will be

glad that I have recommended it, seeing that he pro-
fesses that it is the knowledge of *truth* that he seeks
after.[174]

The terminology, it goes without saying, is very scholastic,
possibly deliberately so in order to be intelligible to the "peripatetics" of the Jesuit college, and is one of Newton's clearest
expressions of the demonstrative methodology that constituted for him the ideal of experimental science.

Despite Newton's reference to "demonstration" and to
having discovered the "true cause" of the elongation of the
image, he does not, to my knowledge, discuss the precise
cause he had used as the middle term in the demonstration,
nor does he indicate to which of the four Aristotelian species
of causality the "true cause" would pertain. Quite obviously
he had in mind some type of material causality, in the sense
of the kind of rays out of which light is composed. Thus he
felt that he had demonstrated a quantitative modality of an
effect through the material parts, or the components, of the
cause that produces it. Newton was, of course, insistent that
the rays of colored light are present in white light, and thus
that they are in some way its component parts, into which it
is resolved by the diffracting action of the prism. Yet precisely how the component rays are present in white light is
one of the questions to which Newton refused to give an answer, despite the numerous suggestions of his critics and adversaries. In this connection, it has recently been proposed
that Newton was secretly committed to a corpuscular theory
of light, and that a suppressed premise stating this is necessary to make sense of his insistence on the demonstrative
force of his argument.[175] It would seem, however, that this interpretation is not the only one possible; anyone with a good
scholastic background, and particularly the Jesuits to whom
Newton addressed his clarificatory remarks, should have been
able to discern a number of possibilities. They might have answered, for example, that the component rays of colored light
are present in white light, not actually, nor potentially, but
merely virtually, after the analogy of the ways in which elements were thought by Aquinas to be present in compounds.
Whether such a latitude of interpretation was explicitly

known to Newton is, of course, impossible to ascertain, although it is noteworthy that Newton never seems to state that the rays contained in white light are themselves actually colored, but rather that they are "disposed to exhibit" a certain color, which would seem to indicate a certain virtuality of their presence in white light.[176]

b. The Cause of Gravity

A yet more fascinating study in Newton is the development of his thought on the nature and cause of gravity, a topic about which he speculated much, and where his conclusions bear a certain resemblance to those already seen in relation to light and color.[177] Since space does not permit a detailed presentation of this topic, we shall have to content ourselves with some summary remarks based on Motte's translation of the *Mathematical Principles of Natural Philosophy*.[178] Here, at the very outset of Book III, where Newton is to treat of the system of the world as it can be ascertained from the propositions of the preceding books, he states his four famous "Rules of Reasoning in Philosophy," rules which Whewell insisted he had derived mainly to establish the existence of gravity.[179] After enumerating these rules, Newton gives an explanation of each — rather extensive for the third but quite brief for the others. The rules and their explanations read as follows: [180]

RULE 1. *We are to admit no more causes of natural things than such as are both true and sufficient to explain their appearances.*

To this purpose the philosophers say that Nature does nothing in vain, and more is in vain when less will serve; for Nature is pleased with simplicity, and affects not the pomp of superfluous causes.

RULE 2. *Therefore to the same natural effects we must, as far as possible, assign the same causes.*

As to respiration in a man and in a beast; the descent of stones in *Europe* and in *America;* the light of our culinary fire and of the sun; the reflection of light in the earth and in the planets.

Rule 3. *The qualities of bodies, which admit neither in-tensification nor remission of degrees, and which are found to belong to all bodies within the reach of our ex-periments, are to be esteemed the universal qualities of all bodies whatsoever.*

For since the qualities of bodies are only known to us by experiments, we are to hold for universal all such as universally agree with experiments; and such as are not liable to diminution can never be quite taken away. . . .

Rule 4. *In experimental philosophy we are to look upon propositions inferred by general induction from phenom-ena as accurately or very nearly true, notwithstanding any contrary hypotheses that may be imagined, till such time as phenomena occur, by which they may either be made more accurate, or liable to exceptions.*

This rule we must follow that the argument of in-duction may not be evaded by hypotheses.[181]

We see from these rules that Newton conceived scientific method as essentially a search for causes. Also underlying his method is his firm belief in the uniformity of nature, which en-ables him to employ inductive argumentation, and thus gener-alize on the basis of limited but careful observation and ex-perimentation.

The examples Newton offers in justification of Rule 2 are of interest, both in themselves and for their application in the more extended comment on Rule 3. The reference to "respira-tion in a man and in a beast" recalls Harvey's studies in com-parative anatomy and the validity of his experimentation on some eighty species of animals to verify the circulation of the blood in man. The example of "the descent of stones in Eu-rope and America" extends the validity of Newton's gravita-tional theories beyond the British Isles to the shores of the new continent then being developed, with the implicit asser-tion that, when one travels to America, he need not start a new science of falling objects because he now experiences stones falling in a different place. Again, "the light of our cu-linary fire and of the sun" suggests a still more extended ap-plication of scientific principles, so that if one can explain how combustion takes place on earth, and does so correctly, he will also understand how the processes of combustion take place on the surface of the sun. "The reflection of light in the earth, and in the planets" is a similar example, for this opens the way to the extension of the science of optics, as learned on earth, to the entire solar system. It is in this way that Newton conceived causal explanation, when obtained from effects studied close at hand and with the aid of experiments, as enabling one to obtain scientific knowledge of even the most remote parts of the universe.

From what has been said it should be obvious that Newton's approach to nature, while heavily mathematical, was also strongly empirical, in the sense that he was convinced that man can only come to a knowledge of the properties of bodies through an examination of their observable characteristics.[182] His classical reasoning based on this conviction occurs in the explanation of his third rule, where he justifies the extension of the principle of "mutual gravitation" to all the bodies of the solar system, and this line of argument now merits a brief analysis.

After his opening statement that "the qualities of bodies are only known to us by experiments" — note the medieval language of "intensification" and "remission" — and various examples by which he shows that sense knowledge and experience are the basis for our assigning attributes to all bodies, Newton cites the experimental evidence that has led him to the law of universal gravitation. Since the empirical data of astronomy show that a uniformly accelerated type of motion does actually occur in the heavens, the inference inescapably follows, by a posteriori demonstration, that celestial matter is no different from terrestrial matter in the sense that both matters undergo a "falling" motion. This, of course, is only another way of saying that all observable bodies gravitate, and thus are endowed with a gravitational principle; or, to put it in yet another way, all observable bodies in the universe have gravity, or are heavy. No longer can one maintain, therefore, that some bodies are essentially light or are composed of a quintessence whose natural motion is eternally circular. Newton realizes that his evidence for this momentous conclusion is not itself overwhelming, and yet he believes it to be sufficient in the light of his third rule. So he concludes, "If it universally appears, by experiments and as-tronomical observations," that all bodies we can see gravitate, "we must, in consequence of this rule, universally allow that all bodies whatsoever are endowed with a principle of mutual gravitation." [183]

In a paragraph immediately preceding this statement, Newton had gone through a similar justification as to why we believe that all bodies are endowed with "extension, hard-ness, impenetrability, mobility, and inertia," and states that the process by which we arrive at such generalizations "is the foundation of all philosophy." [184] Reassured by this analogy, Newton feels that he has now a solid basis for extending the principles established in the first two books of the *Principia* to demonstrate the system of the world, to show that it is he-liocentric precisely in terms of the law of universal gravita-tion, and thus that the sun and the planets do constitute a true solar system. Thus here, as in his optical researches, Newton employs causal reasoning and gives demonstrative force to his arguments by invoking the principle of induction. He is fearful, again, of hypotheses, particularly those that "may be imagined" as alternative but contrary explanations, and explicitly adds his fourth rule to safeguard against any attempt of this type to vitiate demonstrative argument or the inductive process on which it must be based.[185]

Newton's reasoning with regard to gravity and its nature bears a significant resemblance to his reasoning regarding the nature of light and colors, and it encountered much the same type of opposition from his critics. With regard to the system of the world, he was convinced that he had shown gravity to be physical and real, and existing in the planets and their sat-ellites, as well as in the sun, earth, and moon, all of which "are endowed" with a gravitational principle.[186] Thus a planet is maintained in its elliptical orbit around the sun by two forces, one deriving from its inertia, which urges it to fly off into space tangentially in a straight line, and the other de-riving from its gravity, which urges it toward a central body, the sun, by a type of centripetal attraction. For Newton, if the planet's motion is real, if the momentum that would carry it off into space is real, then its physical attraction to the cen-ter of the solar system must be real also. This he believed he had demonstrated, and so he could state with assurance that he had "explained the phenomena of the heavens and of our sea by the power of gravity," [187] just as he had maintained that he had discovered the "true cause" of the elongation of the spectrum by the composition of white light. When pressed further, however, and asked what the cause of this power of gravity might be, and particularly when he saw that the query might involve him in some type of mechanistic hypothesis, Newton's answer was the same as that concerning the ultimate nature of light. Just as he did not pretend to know the nature of light, so he has "not been able to discover the cause of those properties of gravity," and in this regard he would prefer not to feign any hypotheses.[188] This is the context in which occurs the statement with which we began this chapter, and it tempers considerably Newton's so-called agnosticism with regard to causal explanations. The full text in which he explains his position on this difficult question is the following:

> Hitherto we have explained the phenomena of the heavens and of our sea by the power of gravity, but have not yet assigned the cause of this power. This is certain, that it must proceed from a cause that penetrates to the very centres of the sun and planets, without suffering the least diminution of its force; that operates not according to the quantity of the surfaces of the particles upon which it acts (as mechanical causes used to do), but according to the quantity of the solid matter which they contain, and propagates its virtue on all sides to immense distances, decreasing always as the inverse square of the distances. . . . But hitherto I have not been able to discover the cause of those properties of gravity from phenomena, and I feign no hypotheses; for whatever is not deduced from the phenomena is to be called an hypothesis; and hypotheses, whether metaphysical or physical, whether of occult qualities or mechanical, have no place in experimental philosophy. In this philosophy particular propositions are inferred from the phenomena, and afterwards rendered general by induction. Thus it was that the impenetrability, the mobility, and the impulsive force of bodies, and the laws of motion and of gravitation, were discovered. And to us it is enough that gravity does

really exist, and act according to the laws which we have explained, and abundantly serves to account for all the motions of the celestial bodies, and of our sea.[189]

The "hypotheses" or "mechanical causes" to which Newton refused to commit himself, as Koyré has so brilliantly shown, were the Cartesian explanations of gravity through vortex motions in the ether, or other mechanistic explanations such as those proposed by Hooke, Huygens, and Leibniz.[190] For example, Hooke, after making several efforts to explain this "gravitating power," [191] settled on quick vibrations of the ether in which he thought the planets were immersed as the proper mechanical explanation. Huygens set himself more explicitly to defend the Cartesian vortices, and in his *Discourse on the Cause of Gravity* of 1690, argued that gravity is not a property existing within bodies but is rather the effect of an extraneous action.[192] Again, Leibniz, in his *Tentamen de motuum caelestium causis* of 1687, declared himself an admirer of Kepler's views on the causes of gravity, but refused to admit that it was a property of bodies. In his own words, "I think that there remains nothing else but to admit that the celestial motions are caused by the motions of the ether, or, to speak astronomically, by [the motions of] the deferent orbs [which are, however, not solid] but fluid." [193]

By a strange type of irony, just as Newton was suspected of falling back into peripatetic explanations of the nature of light, so he was accused of invoking gravity as an "occult quality," with all the shortcomings of medieval explanations.[194] His adversaries, of course, were not adverse to causal explanations, since their own mechanical hypotheses were as causal as any explanations could be. Their precise difficulty with Newton was that they felt he was falling back on "occult causes," whereas they were in possession of the "true causes" of gravitational and other phenomena. Who was the more correct — itself a most interesting question — need not be decided here. What is inescapable is the conclusion that causality was far from being a dead issue with the founders of modern science. If anything, in fact, it served for them as a touchstone in terms of which they sought to test the truth or falsity of any explanation, and in this sense was an integral component of their scientific methodology.

Notes

CHAPTER ONE

1. New York: Harcourt, Brace and World.

2. P. 15; see also pp. 26–28. Similar statements are made by Carl G. Hempel and Paul Oppenheim in their classical essay, "Studies in the Logic of Explanation," *Philosophy of Science*, 15 (1948), pp. 135–175; significant portions of this are reprinted in Baruch A. Brody, ed., *Readings in the Philosophy of Science*, Englewood Cliffs: Prentice-Hall, 1970, pp. 8–27. Other essays edited by Brody that make comparable assertions are Sylvain Bromberger, "Why-Questions?", pp. 66–87, and Michael Scriven, "Explanations, Predictions, and Laws," pp. 88–104.

3. A negative answer to this question would seem to be implied in T. S. Kuhn's influential book, *The Structure of Scientific Revolutions*, Chicago: University Press, 1962; 2d enlarged ed., 1970.

4. For the documentary sources, see *The Principle of Relativity: A Collection of Original Memoirs on the Special and General Theory of Relativity*, by H. A. Lorentz, A. Einstein, H. Minkowski and H. Weyl, with notes by A. Sommerfeld, tr. by W. Perrett and G. B. Jeffery, London: Methuen, 1923; reprinted Dover, n.d. Note Minkowski's prophecy, made in 1908: "Henceforth space by itself, and time by itself, are doomed to fade into mere shadows, and only a kind of union of the two will preserve an independent reality." — p. 75.

5. A readable account is to be found in Werner Heisenberg, *The Physicist's Conception of Nature*, tr. by A. J. Pomerans from *Das Naturbild der heutigen Physik* (1955), London: Hutchinson, 1958, pp. 38–46.

6. This school has been closely identified with the Vienna Circle and, in the U.S., with the Unified Science movement. The more important papers of the latter group have recently appeared under the title, *Foundations of the Unity of Science*, ed. Otto Neurath *et al.*, 2 vol., Chicago: University Press, 1970.

7. For book-length treatments, see Mario Bunge, *Causality: The Place of the Causal Principle in Modern Science*, Cambridge, Mass.: Harvard University Press, 1959; David Bohm, *Causality and Chance in Modern Physics*, New York: Van Nostrand, 1957; V. F. Lenzen, *Causality in Natural Science*, Springfield, Ill.: Charles Thomas, 1954; Max Born, *Natural Philosophy of Cause and Chance*, Oxford: University Press, 1949.

8. Typical of the interest in realism and metaphysics are: Rom Harré, *The Principles of Scientific Thinking*, Chicago: University Press, 1970; *Matter and Method*, New York: St. Martin's Press, 1964; and *Theories and Things*, London: Sheed & Ward, 1961; Mary B. Hesse, *Models and Analogies in Science*, Notre Dame: University Press, 1966, and *Forces and Fields*, London: Nelson, 1961; Gerd Buchdahl, *Metaphysics and the Philosophy of Science. The Classical Origins: Descartes to Kant*. Cambridge, Mass.: M.I.T. Press, 1969; and L. K. Nash, *The Nature of the Natural Sciences*, Boston: Little, Brown & Co., 1963. Historians of science such as Alexandre Koyré, in his *Metaphysics and Measurement. Essays in the Scientific Revolution*. Cambridge, Mass.: Harvard University Press, 1968, and *Newtonian Studies*, Cambridge, Mass.: Harvard University Press, 1965, have also stimulated a return to more traditional philosophical interests.

9. This is true generally of Nagel, *op. cit.*, and much of the periodical literature; for a survey of the bibliography, see M. W. Wartofsky, *Conceptual Foundations of Scientific Thought: An Introduction to the Philosophy of Science*, New York: Macmillan, 1968, pp. 519–527.

10. Points similar to these have been made by E. A. Burtt, *The Metaphysical Foundations of Modern Science*, 2d rev. ed., New York: Humanities, 1932; and by S. L. Jaki, *The Relevance of Physics*, Chicago: University Press, 1966, and in his introduction to Pierre Duhem's *To Save the Phenomena*, tr. by E. Doland and C. Maschler, Chicago: University Press, 1969, pp. ix–xxvi.

11. Pioneer studies were those of Pierre Duhem, *Études sur Léonard de Vinci*, 3 vol., Paris: Hermann, 1906–1913 and of Anneliese Maier, *Die Vorläufer Galileis im 14. Jahrhundert*, Rome: Edizione di Storia e Letteratura, 1949; see also A. C. Crombie, *Robert Grosseteste and the Origins of Experimental Science*, Oxford: University Press, 1953.

12. One has to be careful here in guarding against a naive view of the influence of any methodological treatise on actual scientific work; for some of the necessary qualifications, see Koyré's review of Crombie's work on Grosseteste, *Diogenes* 16 (1956), pp. 13 ff.

13. Bk. I, ch. 2, 71b8–12. In the translation of G. R. G. Mure: "We suppose ourselves to possess unqualified scientific knowledge of a thing . . . when we think that we know the cause on which the fact depends, as the cause of that fact and no other, and, further, that the fact could not be other than it is."

14. Bk. II, ch. 1, 89b21–25. In his English translation of this passage, Mure expands the text to read as follows: "The kinds of question we ask are as many as the kinds of things which we know. They are in fact four: (1) whether the connexion of an attribute with a thing is a fact, (2) what is the reason of the connexion, (3) whether a thing exists, (4) what is the nature of the thing."

15. Bk. II, ch. 2, 90a5–7. For a complete analysis, see Melbourne G. Evans, "Causality and Explanation in the Logic of Aristotle," *Philosophy and Phenomenological Research*, 19 (1958–59), pp. 466–485.

16. These texts wherein Aristotle exposes the relationships between causality and scientific explanation are brief and hardly intelligible to a modern-day reader. The particular difficulty of the *Posterior Analytics*, in fact, had been emphasized even in the tradition of the Greek commentators, for in the fourth century A.D. Themistius called attention to its many obscurities. Two centuries later Boethius claims to have translated the work into Latin, but his version is no longer extant, although the otherwise anonymous translator, John, acknowledges having seen a partial and corrupt version of the Boethius text in the twelfth century. Some time before 1159, however, the Greek text was translated anew into Latin by James the Venetian-Greek, and this became the *versio vulgata* of the Middle Ages—used by both Robert Grosseteste and Albertus Magnus and often regarded, though erroneously, as the work of Boethius. James's translation is quite accurate, and cryptic on that account, so that there were few teachers who were willing to expound the text. John of Salisbury observes in his *Metalogicon* that "there are almost as many obstacles [to understanding] as there are chapters," attributing this to the poor manuscript tradition. In 1159 John (the anonymous translator) wrote disparagingly of James's version, which he himself did not have, and newly translated the work from the Greek himself. This translation, now available in the *Aristoteles Latinus*, seems to have been little known in the Middle Ages. At about the same time as John was working from the Greek text, a third translation was made from an Arabic version by the Italian Gerard of Cremona, who also made available the commentary of Themistius. Gerard's translation was never as popular as James's, but it was widely used in the Middle Ages, by Albertus Magnus among others. Finally, a fourth translation from the Greek was made by the Dominican William of Moerbeke, who essentially revised James's translation, and whose superior version was used by Thomas Aquinas for his commentary, which is one of the more complete expositions of Aristotelian methodology in the high Middle Ages.

Some idea of the differences in these translations may be seen from the way in which the four scientific questions of Aristotle are rendered into Latin by the translators. James the Venetian-Greek quite accurately lists the four questions as (1) *quia,* (2) *propter quid,* (3) *si est,* and (4) *quid est,* and William of Moerbeke leaves this translation as it stands. The translator John, not quite so accurate but intelligible nonetheless, gives the questions as: (1) *quod,* (2) *propter quid,* (3) *an est,* and (4) *quid est.* Gerard of Cremona, on the other hand, shows how interpretation is frequently introduced by translation through a second language, in his case Arabic. He presents the questions as: (1) *an hoc insit huic,* (2) *quare istud insit huic,* (3) *an hoc sit,* and (4) *quid est hoc.* Yet his version is helpful, though not an accurate translation, and it is interesting that Mure paraphrases it when giving his English translation from the Greek (see fn. 14 above). For fuller details, consult the four different versions in *Aristoteles Latinus,* Vol. IV, 1–4, ed. L. Minio-Paluello and B. G. Dod, Bruges-Paris: Desclée de Brouwer, 1968.

17. Bk. I, ch. 13, 78a22.

18. *Ibid.,* 78a29–78b15.

19. *Ibid.,* 79a2–7. This statement, differently interpreted by various commentators, was unparalleled in the stimulation it gave to an adumbration of mathematical physics within the high and late Middle Ages.

20. *Ibid.,* 79a14–16. John Philoponus, a sixth-century Greek commentator, offers two explanations: because such wounds have the greatest area in relation to their perimeters, and because their healing surfaces are farther apart and nature has difficulty joining them.

21. Bk. II, ch. 11, 94a20–23. The four Greek expressions are translated into Latin as follows: [1] James the Venetian-Greek and William of Moerbeke render the formal cause as *quid erat esse,* while John translates it as *quid est esse;* Gerard of Cremona, working from the Arabic, gives *causa secundum quod ipsa est forma rei et eius intentio.* [2] James describes the material cause as *cum hec sunt necesse est hoc esse,* while William gives *cum hec sint necesse est hoc esse,* and John gives *quibus existentibus necesse hoc esse;* Gerard again is quite different, rendering this as *causa secundum quod ipsa est yle, et est cuius esse sequitur esse forme* (note the transcription of the Greek *hulē,* for matter, directly into the Latin text). [3] James, John, and William all describe the efficient cause as *que aliquid primum movit,* while Gerard gives this as *causa secundum quod ipsa est motor propinquus.* [4] James and William translate the final cause as *cuius causa,* John as *cuius gratia,* and Gerard as *causa secundum quod ipsa est causa finalis.*

22. Bk. II, ch. 3, 194b23–195a27. In the Middle Ages, Latin translations of the *Physics* were made from the Greek by James and William also. Michael Scot prepared another translation from

the Arabic, with the intention of making a version available that would not fall under the bans of 1210 and 1215 pronounced at Paris against the reading of Aristotle's books on natural philosophy and their commentaries. As in the case of Gerard of Cremona, Michael's translations differ considerably from those of James and William, but the variations are not significant for what follows. In modern English usage the term "cause" has taken on such different connotations since its employment by Aristotle that it is quite difficult to make his meaning intelligible to the English reader. Four printed translations are now available—those of Wicksteed and Cornford, Hardie and Gaye, Hope, and Apostle—and these show as much or more divergence among themselves as do the three Latin texts. (Account is not taken here of the translation now in process by W. Charlton, of which only the first two books have thus far appeared — *Aristotle's Physics I, II.* New York: Oxford University Press, 1970.)

23. Practically all of these expressions occur in the *Physics*, Bk. II, ch. 3, 194b16–195a1, much of which material is repeated, almost verbatim, in the *Metaphysics*, Bk. V, ch. 2. For details of these occurrences consult the [Greek] Analytical Index of Technical Terms in *Aristotle's Physics*, tr. Richard Hope, Lincoln: University of Nebraska Press, 1961, pp. 180–240; also a similar index in *Aristotle's Metaphysics*, tr. Richard Hope, New York: Columbia University Press, 1952, pp. 319–390.

The Latin translations of these terms are generally unexceptional, although they manifest the nuances that one might expect from the Greek-Latin and Arabic-Latin traditions. [1] The Greek expressions for formal cause are variously rendered as *species, exemplum, ratio ipsius quod quid erat esse, totum, compositio,* and *diffinitio* by James and William, and as *forma, imago, ratio significans quidditatem rei, genus,* and *universum* by Michael Scot. Shape (*schēma*), when employed in the sense of formal causality, and this is rarely, is rendered *figura* by all translators, whereas *idea* is translated as *idea* by James and William and as *forma* by Michael. [2] Expressions for material causality are given as *ex quo fit, materia, subiectum, elementa, partes,* and *suppositiones* by James and William, and as *ex quo fit, materia, quasi-subiectum, partes* and *propositiones* by Michael. [3] For the efficient cause all the translators give *principium primum mutationis aut quietis,* which James and William characterize as a *faciens* and Michael as an *agens.* [4] The final cause is translated as *finis, bonum,* and *propter quid* by all three, and additionally as *cuius causa* and *gratia* by James and William.

Critical editions of these Latin texts are not yet available. For the James of Venice and Michael Scot versions I have used the *editio princeps* of Averroës's commentary on the *Physics* (Padua, *c.* 1472–75), and for the Moerbeke version the Latin text printed with Giles of Rome's commentary on the *Physics* (Venice, 1502) as slightly more accurate than that included in the manual Leonine edition of St. Thomas's commentary (Rome, 1954).

24. *Physics,* Bk. II, ch. 7, 198a12–198b9. The concluding paragraph of this chapter is cryptic, and Hope's translation is more a paraphrase than a literal reading: "In short, the question 'why?' calls for a comprehensive answer. Thus [we must explore (1) the efficient factor:] 'this must result from *that,*' and 'from that' either without qualification or in most cases; [(2) the material factor:] 'if this is to be, then *that* must be,' just as syllogistic conclusions are conditioned by their premises; [(3) the formal factor:] '*that* was what it meant for this to be'; and [(4) the final factor:] '*that* is why it is best for this to be thus and so,' not of course absolutely but relatively to its distinctive being."

25. R. K. Sprague, "The Four Causes: Aristotle's Exposition and Ours," *The Monist* 52(1968), pp. 298–300.

26. G. E. L. Owen, "Aristotle: Method, Physics, and Cosmology," *Dictionary of Scientific Biography,* Vol. I (New York: Scribners, 1970), esp. pp. 254d–255d, 257b–258b. See also Owen's "*Tithenai ta phainomena,*" *Aristote et les problèmes de méthode.* Communications présentées au Symposium Aristotelicum tenu à Louvain du 24 août au 1ᵉʳ septembre 1960. Louvain: Publications Universitaires, 1961, pp. 83–103.

27. L. Minio-Paluello, "Aristotle: Tradition and Influence," *ibid.,* esp. pp. 277a–279c.

28. G. E. L. Owen, *ibid.,* p. 252b.

29. *Ibid.,* esp. pp. 251c, 254c, 256c.

30. In his *Logic of Scientific Discovery,* New York: Basic Books, 1959. Aristotle seems to have been less insistent on the role of verification, or the observational confirmation of theories, although he was not unaware of its existence. In his treatise *On the Generation of Animals,* for example, when discussing the copulation of bees, he concludes: "Such appears to be the truth about the generation of bees, judging from theory and from what are believed to be the facts about them; the facts, however, have not yet been sufficiently grasped; if ever they are, then credit must be given rather to observation than to theories, and to theories only if what they affirm agrees with the observed facts" (Bk. III, ch. 10, 760b27–33). Arthur Platt, who translated this work for the W. D. Ross English edition (Oxford, 1910), notes that the German translation of the sentence by Aubert and Wimmer (Leipzig, 1860) was deservedly printed in conspicuous type and remarks that "it should have been kept in mind by those bastard Aristotelians who at the revival of learning refused to accept observed facts because they were supposed to contradict Aristotle's statements (which they often did not)." Aristotle also makes a tangential reference to observational verification when discussing the concentration of light rays through a burning glass in *Posterior Analytics,* Bk. I, ch. 31, 88a12–17.

31. *Meteorologica,* tr. H. D. P. Lee, Loeb Classical Library, Cambridge, Mass.: Harvard University Press, 1952, p. 49.

32. *Ibid.,* p. 371.

33. For fuller details, see S. Sambursky, *The Physical World of the Greeks,* tr. M. Dagut, New York: Macmillan, 1956, esp. ch. 2.

34. *Metaphysics,* Bk. I, ch. 5, 986a18.

35. In the words of the Pythagorean philosopher Philolaus: "For the nature of Number is the cause of recognition, able to give guidance and teaching to every man in what is puzzling and unknown. For none of existing things would be clear to anyone, either in themselves or in their relationship to one another, unless there existed Number and its essence. But in fact Number, fitting all things in to the soul through sense-perception, makes them recognizable and comparable with one another."—cited by Sambursky, *op. cit.* (Collier Books ed., 1962), pp. 46–47.

36. *Timaeus,* 49B, 50D, 51A.

37. *Ibid.,* 50B–53C.

38. *Ibid.,* 54A–56D. For a detailed analysis, see William Pohle, "The Mathematical Foundations of Plato's Atomic Physics," *Isis,* 62 (1971), pp. 36–46.

39. An early twelfth-century Byzantine poet, John Tzetzes, is the source of the statement that Plato inscribed over the entrance to the Academy, "Let no one ignorant of geometry enter my door."

40. *Republic,* Bk. VII, 527D–530D.

41. *Timaeus,* 29D; also *Republic,* Bk. VII, 530D–533A, and *Philebus,* 55D–57E. For fuller details, see Sambursky, *op. cit.,* pp. 60–63.

42. See J. A. Weisheipl, *The Development of Physical Theory in the Middle Ages,* New York: Sheed and Ward, 1959, pp. 14–17; reprinted Ann Arbor: University of Michigan Press, 1971.

43. Pierre Duhem, *To Save the Phenomena,* p. 5. See also Jürgen Mittelstrass, *Die Rettung der Phänomene: Ursprung und Geschichte eines antiken Forschungsprinzips.* Berlin: Walter de Gruyter, 1962.

CHAPTER TWO

1. For a general characterization of Oxford University and its relationships to the University of Paris during the Middle Ages, see Gordon Leff, *Paris and Oxford Universities in the Thirteenth and Fourteenth Centuries:* An Institutional and Intellectual History, New York: John Wiley & Sons, 1968, esp. pp. 271–309.

2. *Compendium studii philosophie,* Brewer ed., p. 469, cited by L. Baur, *Die Philosophie des Robert Grosseteste, Bischofs von Lincoln.* Beiträge zur Geschichte der Philosophie des Mittelalters, Band XVIII, 4–6, Münster: 1917, p. 15°.

3. Cf. Crombie, *Robert Grosseteste . . . ,* pp. 44–188, and

"Grosseteste's Position in the History of Science," in D. Callus, ed., *Robert Grosseteste: Scholar and Bishop,* Oxford: University Press, 1955, pp. 98–120. Crombie's work is summarized in Leff, *op. cit.,* pp. 272–286; also in Weisheipl, *Development . . . ,* pp. 50–52. See also Bruce S. Eastwood, "Medieval Empiricism: The Case of Grosseteste's Optics," *Speculum* 43 (1968), pp. 306–321, for a slightly different view.

4. Grosseteste translated from the Greek the first two books of *De caelo* and the entire *Nicomachean Ethics,* to which he also added a commentary. He composed a series of theological treatises, the most important of which is entitled *De libero arbitrio,* and a number of Biblical commentaries, of which that on the Hexaëmeron is the best known. When he became Bishop of Lincoln he still continued his interest in Greek letters, and translated, probably with the assistance of others, the entire pseudo-Dionysian corpus with its heavy Proclean content, which he accompanied by commentaries, and works of Greek Fathers such as John Damascene and Maximus the Confessor.

5. Following the summary in Leff, *op. cit.,* pp. 278–285.

6. *De lineis angulis et figuris seu de fractionibus et reflexionibus radiorum,* Baur ed., pp. 59–60; according to Crombie, this work seems to suppose knowledge of the *Posterior Analytics* and was probably written after Grosseteste's commentary on the latter — see *Robert Grosseteste . . . ,* pp. 47–51.

7. Grosseteste's teaching on these matters in his *Physics* commentary, which is not discussed here for reasons of space, is completely consistent with that in the commentary on the *Posterior Analytics.* See *Robert Grosseteste Episcopi Lincolniensis Commentarius in VIII Libros Physicorum Aristotelis,* ed. Richard C. Dales, Boulder: University of Colorado Press, 1963; this work is summarized in the editor's "Robert Grosseteste's *Commentarius in Octo Libros Physicorum Aristotelis,*" *Medievalia et Humanistica* 11 (1957), pp. 10–33.

8. *Roberti Lincolniensis . . . in Posteriorum Analeticorum librum* [sic], Venice: Octavianus Scotus, 1521, Lib. 1, cap. 12, comm. 65, fol. 16va.

9. Fol. 16va–vb. For an English translation, see Crombie, *Robert Grosseteste . . . ,* p. 53.

10. 75b23–24. Grosseteste, Lib. 1, cap. 7, comm. 39, fol. 9vb–10ra. Crombie gives a partial translation of this, but unfortunately it is marred by too many ellipses to be intelligible; see *op. cit.,* pp. 130–131. Etienne Gilson gives the Latin text in his "Pourquoi S. Thomas a critiqué S. Augustin," *Archives d'histoire doctrinale et littéraire du moyen âge* 1 (1926), pp. 95–96 fn., and the teaching is discussed in L. E. Lynch, "The Doctrine of Divine Ideas and Illumination in Robert Grosseteste," *Mediaeval Studies* 3 (1941), pp. 161–173, esp. p. 169; Eastwood gives a summary in "Medieval Empiricism . . . ," pp. 308–309.

11. Cap. 8, comm. 41, fol. 10rb–va, commenting on 75b32–36.

12. *Ibid.*, fol. 10va.

13. The lunar eclipse is discussed in *De sphaera*, Baur ed., p. 29. For a comparison of this work with the *Sphere* of Sacrobosco, see Lynn Thorndike, *The Sphere of Sacrobosco and Its Commentators*, Chicago: University Press, 1949, pp. 10–14.

14. Here Grosseteste seems to anticipate a teaching that was to be more explicitly enunciated by Thomas Aquinas in his commentary. See *infra*, Chap. 3, pp. 75 ff.

15. Cap. 12, comm. 65, fol. 16vb; this portion of the commentary is not translated by Crombie.

16. *Ibid.*

17. . . . forte eiusdem conclusionis est scientia propter quid et scientia quia in eadem scientia — *ibid*.

18. . . . licet Aristoteles de hoc non ponat exemplum — *ibid*.

19. For an explanation of this theory, developed by Grosseteste but perfected by Roger Bacon and John Peckham, see David C. Lindberg, *John Pecham and the Science of Optics. Perspectiva communis*, edited with an introduction, English translation, and critical notes. Madison: University of Wisconsin Press, 1970, pp. 34–39.

20. Fol. 16vb.

21. *Ibid.*

22. On this understanding neither demonstration would be *propter quid*, however, since both would be from effect to cause and hence *quia*.

23. Baur ed., pp. 51–59; Eng. tr. C. C. Riedl, *Robert Grosseteste on Light*, Milwaukee: Marquette University Press, 1942.

24. Cf. Leff, *Paris and Oxford* . . . , p. 283.

25. Cap. 12, comm. 65, fol. 16vb–17ra.

26. *Ibid.*, fol. 16vb.

27. *Ibid.*, fol. 17ra. Thus the moon passes through its crescent and gibbous phases and "when the two circles intersect orthogonally," i.e., between the crescent and the gibbous, the termination of the illumined portion appears "as a straight line" (fol. 16vb).

28. See the discussion of the proofs for the earth's sphericity, *infra*, Chap. 3, pp. 81–86.

29. Cf. Crombie, *Robert Grosseteste* . . . , pp. 91–127.

30. Cap. 12, comm. 67, fol. 17v; most of this comment is translated by Crombie, *ibid.*, pp. 91–93.

31. *Ibid.*; Crombie, p. 92.

32. *Ibid.*; Crombie, p. 93.

33. *Ibid.*

34. See B. S. Eastwood, "Robert Grosseteste's Theory of the Rainbow," *Archives internationales d'histoire des sciences* 78 (1966), pp. 313–332; also Eastwood, "Medieval Empiricism . . . ," pp. 317–320; and Richard C. Dales, "Robert Grosseteste's

Scientific Works," *Isis* 52 (1961), pp. 381–402, esp. pp. 399–401.

35. Baur gives the reading: "Et perspectivi et physici est speculatio de iride. Sed ipsum 'quid' physici est scire, 'propter quid' vero perspectivi. Propter hoc Aristoteles in libro meteorologicorum non manifestavit 'propter quid,' quod est perspectivi, sed ipsum 'quid' de iride, quod est physici, in brevem sermonem coarctavit." — *ed. cit.*, p. 72. In the works just cited, Crombie (p. 117), Dales (p. 399), and Eastwood (p. 307), all follow this reading, saying that geometrical optics gives the *propter quid* of the rainbow whereas physics gives the *quid*. A better reading of the MSS, however, would replace the 'quid' given by Bauer with 'quia'; this squares properly with Aristotle's teaching on the four scientific questions, whereas Baur's reading does not. Another possibility, based on only one MS however, would be to replace the 'quid' by 'quod'; this would be consistent with the Latin translation of Aristotle's four questions given by the translator John. See Chap. 1, *supra*, note 16, toward the end.

36. This is suggested by Eastwood, "Medieval Empiricism . . . ," pp. 308–309.

37. Cap. 8, comm. 42, fol. 10vb–11a.

38. On Bryson, see the article by Philip Merlan, *Dictionary of Scientific Biography*, Vol. 2, pp. 549–550; Grosseteste's discussion of Bryson suggests that he may have incorporated some elements from Archimedes' *Measurement of a Circle*, ed. T. L. Heath, *The Works of Archimedes*, New York: Dover, 1953, pp. 91–93.

39. *Ibid.*, fol. 11ra.

40. *Ibid.*; for an English translation of part of this discussion, see Crombie, *Robert Grosseteste . . .* , pp. 95–96.

41. *Opera omnia*, ed. J. L. Heiberg, Leipzig: 1895, pp. 287–289, proof of Theorem 1; cf. Theorems 7 and 16.

42. *Ibid.*; Crombie, p. 95.

43. *Ibid.*; Crombie, p. 96. Some have seen in this explanation a principle of uniformity of nature. In the *De iride*, when explaining reflection and refraction, Grosseteste enunciates what has also been referred to as a principle of economy (*lex parsimoniae*), namely, "the whole operation of nature is by the most ordered, shortest, and best means possible . . .", Baur ed., p. 75. Both may be regarded as types of teleological explanation. See B. S. Eastwood, "Grosseteste's 'Quantitative' Law of Refraction: A Chapter in the History of Non-experimental Science," *Journal of the History of Ideas* 28 (1967), pp. 403–414, esp. pp. 406–407, 412; also "Medieval Empiricism . . . ," p. 318.

44. Cap. 12, comm. 67, fol. 17vb.

45. 79a14–16.

46. Fol. 17vb.

47. *Ibid.*

48. Fol. 17va; Crombie, p. 92.

49. See Dales, "Robert Grosseteste's . . . ," pp. 392–393.

50. Aristotle, *De sophisticis elenchis*, chap. 5, 167b1–20.

51. Crombie, p. 82 fn. 5, cites Grosseteste's use of the expression "per modum verificationis" in cap. 1, comm. 1, fol. 2rb, but in its context this has no connection with the methodological problem under discussion.

52. See *supra*, Chap. 1, note 30.

53. Cap. 14, comm. 81, fol. 20vb–21ra; Crombie, pp. 73–74.

54. *Ibid.*, fol. 21ra; Crombie, p. 74.

55. For details, see Eastwood, "Medieval Empiricism . . . ," p. 310.

56. *De iride*, Baur ed., pp. 74–75.

57. *Robert Grosseteste . . . ,* p. 124.

58. Cf. Colin M. Turbayne, "Grosseteste and an Ancient Optical Principle," *Isis* 50 (1959), pp. 467–472, and Eastwood, "Grosseteste's Quantitative' Law . . . ," pp. 408–414.

59. Eastwood, "Medieval Empiricism . . . ," p. 321.

60. Lib. II, cap. 2, comm. 45, fol. 40rb–41va.

61. *Ibid.*, fol. 41va.

62. Bk. II, chap. 9, 369a10–370a35.

63. Comm. 46, fol. 40vb–41ra.

64. *Ibid.*, fol. 41ra: "Ordo autem verborum Aristotelis hic perturbatus est. Sed credo quod praedicto modo debet ordinari ut exponatur eius intentio."

65. Comm. 47, fol. 41ra–rb.

66. *Ibid.*, fol. 41rb.

67. *Ibid.*

68. Cap. 3, comm. 54, fol. 42vb.

69. There is yet more to Grosseteste's analysis, for in cap. 4, comm. 75, fol. 47vb–48ra, he has a detailed discussion of the generation of sound, similar to that in his opuscula *De generatione sonorum* and *De artibus liberalibus;* see Baur, pp. 58°–60°; also Dales, "Robert Grosseteste's . . . ," pp. 383–384.

70. *Commentary on the Posterior Analytics of Aristotle*, tr. F. R. Larcher, Albany: Magi Books, 1970, p. 191.

71. For details see the article on Bacon by A. C. Crombie and J. D. North in the *Dictionary of Scientific Biography*, Vol. 1, pp. 377–385.

72. A biographical sketch of Peckham and a summary of his work is given in Lindberg, *John Pecham . . . ,* pp. 3–51.

73. *The 'Opus Majus' of Roger Bacon*, ed. J. H. Bridges, 2 vols., Oxford: Clarendon Press, 1897, dist. 2, cap. 1, Vol. 1, p. 109; Eng. tr. R. B. Burke, 2 vols., Philadelphia: University of Pennsylvania Press, 1928, Vol. 1, p. 128.

74. *Ibid.*, Bridges ed., p. 110; Burke tr., pp. 129–130.

75. *Ibid.*, Bridges ed., p. 111; Burke tr., p. 130.

76. Part 2, dist. 3, cap. 3, Vol. 2 of Bridges ed., pp. 106–108; Burke tr., Vol. 2, pp. 523–525.

77. Cap. 7, Bridges ed., pp. 120–126; Burke tr., pp. 535–542.

78. *Ibid.*, Bridges ed., p. 120; Burke tr., p. 536; see also Part VI, *Scientia Experimentalis*, cap. 5, Bridges ed., p. 179; Burke tr., p. 594.

79. Cap. 4, Bridges ed., pp. 108–114; Burke tr., pp. 525–530.

80. Part 3, dist. 1–2, Bridges ed., pp. 130–166; Burke tr., pp. 546–575.

81. Cap. 1, Bridges ed., p. 167; Burke tr., p. 583.

82. *Ibid.*, Bridges ed., pp. 167–168; Burke tr., p. 583.

83. *Ibid.*, Bridges ed., p. 168; Burke tr., p. 583.

84. Cap. 12, Bridges ed., pp. 202–203; Burke tr., p. 616.

85. For the relation between Bacon and Peter of Maricourt, see *infra*, Chap. 3, pp. 88 ff.

86. Grosseteste was the first master of the Franciscans at Oxford and had a marked influence on their tradition; for details, see Lindberg, *John Pecham . . .* , pp. 3–11.

87. Preface, Lindberg ed., p. 61.

88. For a summary of the scope and principal ideas of the *Perspectiva communis*, see Lindberg, *John Pecham . . .* , pp. 33–51.

89. "Roger Bacon's Theory of the Rainbow: Progress or Regress?" *Isis* 57 (1966), pp. 235–248.

90. D. C. Lindberg, "The Cause of Refraction in Medieval Optics," *The British Journal for the History of Science* 4 (1968), pp. 23–38; see also Lindberg's "Lines of Influence in Thirteenth-Century Optics: Bacon, Witelo, and Pecham," *Speculum* 46 (1971), pp. 66–83.

91. D. C. Lindberg, "The Theory of Pinhole Images from Antiquity to the Thirteenth Century," *Archive for History of Exact Sciences* 5 (1968), pp. 154–176; "A Reconsideration of Roger Bacon's Theory of Pinhole Images," *ibid.*, 6 (1970), pp. 214–223, esp. p. 216; "The Theory of Pinhole Images in the Fourteenth Century," *ibid.*, 6 (1970), pp. 299–325. Note, in the second article, Peckham's causal terminology when referring to Bacon's contribution: "Others more subtly investigating the cause [of the roundness in radiation that has passed through a triangular aperture] consider the roundness of the sun as a remote cause . . . and the intersection of the rays as the proximate cause." (p. 222)

92. This despite their careful investigation of pinhole images, which cannot compare with the experimental work on the rainbow done only a few decades later by Theodoric of Freiberg. See *infra*, Chap. 3, pp. 94–103.

93. A good summary of the Mertonian contribution and its relation to Ockham and his thought is to be found in J. A. Weisheipl, "Ockham and Some Mertonians," *Mediaeval Studies* 30 (1968), pp. 163–213.

94. For a brief survey of Ockham's life and thought, with texts, see William of Ockham, *Philosophical Writings: A Selection,*

ed. Philotheus Boehner, Edinburgh: Thomas Nelson, 1957; New York: Bobbs-Merrill, 1964.

95. *Ibid.*, pp. xvi–xxiii.

96. *Summa totius logicae,* pars 1, cap. 49, ed. P. Boehner, St. Bonaventure, N.Y.: Franciscan Institute, 1951, p. 141; cited and analyzed by J. A. Weisheipl, "The Place of John Dumbleton in the Merton School," *Isis* 50 (1959), pp. 443–445. See also Weisheipl's *The Development . . .* , pp. 64–68.

97. *Summa totius logicae,* pars 1, cap. 44, in Boehner, *Philosophical Writings,* p. 154.

98. *Tractatus de successivis,* ed. P. Boehner, St. Bonaventure, N.Y.: Franciscan Institute, 1944, cited and analyzed by Herman Shapiro, *Motion, Time and Place According to William Ockham,* St. Bonaventure, N.Y.: Franciscan Institute, 1957, pp. 36–44, esp. p. 40; but cf. Weisheipl, "The Place of John Dumbleton . . . ," pp. 444–445, esp. fn. 31.

99. *Reportatio,* II, q. 26, in Boehner, *Philosophical Writings,* p. 156; cf. Shapiro, *op. cit.,* p. 53.

100. E.g., Sir Edmund Whittaker, *Space and Spirit,* Edinburgh: Thomas Nelson, 1946, pp. 139–143; E. J. Dijksterhuis, *The Mechanization of the World Picture,* tr. C. Dikshoorn, Oxford: University Press, 1961, p. 176.

101. Weisheipl, "Ockham and Some Mertonians," p. 164, fn. 4.

102. On Burley, see the article by John Murdoch and Edith Sylla in the *Dictionary of Scientific Biography,* Vol. 2, pp. 608–612; also Weisheipl, "Ockham and Some Mertonians," pp. 174–188, and "Repertorium Mertonense," *Mediaeval Studies* 31 (1969), pp. 185–208.

103. The two treatises on the intension and remission of forms are entitled *Tractatus primus sive tractatus de activitate, unitate et augmento formarum activarum habentium contraria suscipientium magis et minus* and *Tractatus secundus sive tractatus de intensione et remissione formarum;* for details, see Murdoch and Sylla, *loc. cit.,* p. 611.

104. Burley's teaching is summarized, *ibid.,* p. 610.

105. See Dijksterhuis, *Mechanization . . .* , pp. 174–175.

106. See Murdoch and Sylla, *loc. cit.,* pp. 610–611.

107. According to Weisheipl, "Thomas Bradwardine has rightly been called the 'founder of the Merton School,' even though he was promulgating and developing the scientific ideals defended by Grosseteste, Kilwardby, and Roger Bacon in the thirteenth century." — "Ockham and Some Mertonians," p. 189.

108. Again see Weisheipl, "Ockham and Some Mertonians," pp. 189–195, esp. p. 194.

109. For a careful presentation of the central ideas in Bradwardine's works, see J. E. Murdoch's article in the *Dictionary of Scientific Biography,* Vol. 2, pp. 390–397.

110. "The Place of John Dumbleton . . . ," pp. 447–448.

111. For details, see Weisheipl, *ibid.*, pp. 449–454.

112. Leff, *Paris and Oxford Universities . . .* , p. 303.

113. *Ibid.*

114. Curtis Wilson, *William Heytesbury: Medieval Logic and the Rise of Mathematical Physics,* Madison: University of Wisconsin Press, 1960, p. 24.

115. For the basic texts and commentary, see Marshall Clagett, *The Science of Mechanics in the Middle Ages,* Madison: University of Wisconsin Press, 1959, esp. pp. 235–242, 255–289.

116. There is as yet no comprehensive treatment of Swineshead and his thought; a summary of his *Liber calculationum* is provided by Lynn Thorndike, *A History of Magic and Experimental Science,* 8 vols., New York: Columbia University Press, 1923–1958; Vol. 3 (1934), pp. 370–385, and a significant excerpt from the work is to be found in Clagett, *The Science of Mechanics . . .* , pp. 290–302. See also Carl B. Boyer, *The History of the Calculus and Its Conceptual Development,* New York: Dover Publications, 1959, pp. 69–80, and the article cited in the following note.

117. This occurs in Tractate 11 of the *Liber calculationum,* which is edited, translated, and analyzed in M. A. Hoskin and A. G. Molland, "Swineshead on Falling Bodies: An Example of Fourteenth-Century Physics," *British Journal for the History of Science* 3 (1966), pp. 150–182.

118. J. E. Murdoch, "*Mathesis in philosophiam scholasticam introducta:* The Rise and Development of the Application of Mathematics in Fourteenth-Century Philosophy and Theology," *Arts libéraux et philosophie au moyen âge.* Actes du quatrième congrès international de philosophie médiévale, 1967. Montreal: Institut d'Etudes Médiévales, 1969, p. 231. To this lengthy and excellent article Murdoch appends a rigorous mathematical analysis of Swineshead's argument, pp. 250–254.

119. Hoskin and Molland, "Swineshead . . . ," p. 154.

120. Cambridge, Gonville and Caius MS 499/268, fol. 212ra–213rb.

121. Weisheipl, *The Development . . .* , p. 76; for more details, see W. A. Wallace, "Mechanics from Bradwardine to Galileo," *Journal of the History of Ideas* 32 (1971), pp. 18–19, fn. 14.

122. "The Place of John Dumbleton . . . ," p. 448.

123. "Mechanics from Bradwardine to Galileo," pp. 15–28, esp. pp. 20–21. See also Edward Grant, "Hypotheses in Late Medieval and Early Modern Science," *Daedalus,* Proceedings of the American Academy of Arts and Sciences, 91 (1962), pp. 599–616. On pp. 604–605 Grant states: "There is little doubt that fourteenth-century scientific literature reveals an abundant display of scientific imagination, as evidenced by the frequent occurrence of the phrase *secundum imaginationem* — especially in treatises writ-

ten by nominalists. Significantly, however, these imaginative scientific proposals were formulated for the most part without regard for application to the physical world. The mind was free to create — subject only to the demands of logical consistency. Boldness in scientific speculation was counterbalanced by a noticeable restraint in claiming physical truth for these speculations. The truth or falsity of assumptions or hypotheses was not at issue."

CHAPTER THREE

1. See Leff, *Paris and Oxford Universities* . . . , pp. 188–189.

2. For a survey of Albert's life and teachings, see the article on him by J. A. Weisheipl, *New Catholic Encyclopedia*, New York: McGraw-Hill, 1967, Vol. 1, pp. 254–258.

3. This was a source of annoyance to Roger Bacon, who complained of Albert's having already attained as much authority in the schools as Aristotle, Avicenna, and Averroës; see *Opus tertium*, cap. 9, Brewer ed., p. 30. For further documentation, see J. A. Weisheipl, "Albertus Magnus and the Oxford Platonists," *Proceedings of the American Catholic Philosophical Association* 32 (1958), pp. 124–139, esp. p. 126 and fns. 11–12.

4. *Opera omnia*, ed. A. Borgnet, 38 vols., Paris: Ludovicus Vivès, 1890–1899, Vol. 2, pp. 63, 88, 93, 106, 186, and 198.

5. *Liber I Posteriorum Analyticorum*, tr. 1, cap. 2, Borgnet ed., Vol. 2, p. 5, although it is possible that Albert took the example directly from Galen or Avicenna; see *supra*, p. 42.

6. *Ibid.*, tr. 2, cap. 17, pp. 63–64.

7. *Ibid.*, p. 65.

8. *Ibid.*, tr. 3, cap. 7, p. 86.

9. Lib. II, tr. 2, cap. 12, pp. 194–196. He does treat the causes of thunder and lightning at great length, however, in his exposition of the *Libri Meteororum*, Lib. III, tr. 3, Borgnet ed., Vol. 4, pp. 639–665.

10. *Liber III Physicorum*, tr. 1, cap. 1, Borgnet ed., Vol. 3, pp. 186–187; for an analysis of Albert's teaching, see Anneliese Maier, *Die Vorläufer Galileis im 14. Jahrhundert*, Rome: Edizioni di Storia e Letteratura, 1949, pp. 9–25, esp. pp. 11–16.

11. *Liber I Metaphysicorum*, tr. 1, c. 1, Borgnet ed., Vol. 6, p. 2. This work is now available in the Cologne critical edition, Vol. 16, but there are no important textual differences between the two editions bearing on the topics we will discuss.

12. *Ibid.*, cap. 8, p. 15; see Weisheipl, "Albertus Magnus . . . ," pp. 131–136.

13. Kilwardby had an illustrious career also, serving later as Archbishop of Canterbury, where he condemned Averroistic (and Thomistic) theses at Oxford on March 18, 1277, only ten days after Étienne Tempier's celebrated condemnation at Paris. For some details, see Leff, *Paris and Oxford* . . . , pp. 290–293.

14. This work is described by D. E. Sharp, "The *De ortu scientiarum* of Robert Kilwardby (d. 1279)," *The New Scholasticism* 8 (1934), pp. 1–30; the theory of science contained therein is analyzed by Weisheipl, with citation of texts, in "Albertus Magnus . . . ," pp. 132–135.

15. Cf. Weisheipl, *The Development . . . ,* pp. 53–54.

16. *Liber I Meteororum,* tr. 1, cap. 1, Borgnet ed., Vol. 6, p. 2.

17. Weisheipl, "Albertus Magnus . . . ," pp. 136–137.

18. See my article on Albert in the *Dictionary of Scientific Biography,* Vol. 1, pp. 99–103.

19. *Liber I de Caelo et Mundo,* tr. 4, cap. 10, Borgnet ed., Vol. 4, p. 120.

20. Lib. II, tr. 2, cap. 1, Borgnet ed., Vol. 5, p. 30.

21. Lib. III, tr. 4, cap. 11, Borgnet ed., Vol. 4, p. 679.

22. *Liber VIII Physicorum,* tr. 1, cap. 14, Borgnet ed., Vol. 3, p. 553.

23. *Ibid.,* tr. 2, cap. 2, p. 564.

24. *Ethica,* Lib. VI, tr. 2, cap. 25, Borgnet ed., Vol. 7, pp. 442–443.

25. *Ibid.,* p. 443.

26. See, for example, the illustrations given by Lynn Thorndike, *A History of Magic . . . ,* Vol. 2, pp. 517–592, esp. pp. 546–547.

27. For details, consult A. C. Crombie, *Medieval and Early Modern Science,* 2 vols., New York: Doubleday Anchor, 1959 (rev. ed. of *Augustine to Galileo,* Cambridge, Mass.: Harvard U. P., 1953), Vol. 1, pp. 147–161.

28. Particularly noteworthy in this regard are Ulrich of Strassburg, Giles of Lessines, and Theodoric of Freiberg, the last of whom is discussed in some detail below, pp. 94–103.

29. For further biographical data, see my article on Aquinas in the *Dictionary of Scientific Biography,* Vol. 1, pp. 196–200.

30. Aquinas was also concerned with problems of textual analysis and with the accuracy of the translation on which he was commenting. For an example, see Bk. 2, lect. 8 [n. 3], in Thomas Aquinas, *Commentary on the Posterior Analytics of Aristotle,* tr. F. R. Larcher, Albany: Magi Books, 1970, p. 193.

31. *Ibid.,* Bk. 1, lect. 16, Larcher tr., p. 53; see also Bk. 2, lect. 8, pp. 192–193.

32. Bk. 1, lect. 16, p. 54.

33. *Ibid.*

34. *Ibid.*

35. *Ibid.*

36. *Ibid.,* pp. 54–55, italics mine.

37. See *supra,* p. 31 and p. 67.

38. *Commentary on Aristotle's Physics,* tr. R. J. Blackwell *et al.,* New Haven: Yale U. P., 1963, Bk. 2, lect. 15, nos. 270–272;

Commentary on the Posterior Analytics of Aristotle, Bk. 2, lect. 7, Larcher tr., p. 188; lect. 9, p. 200.

39. 87b18–19.

40. Bk. 1, lect. 42, Larcher tr., p. 148.

41. Even among the medievals, however, the expression *lex naturae* was coming to be applied to the regularities of reflection and refraction and other natural phenomena; see Roger Bacon, *Opus tertium,* Duhem ed., pp. 78, 90; *Opus maius,* Bridges ed., Vol. 2, p. 49; Vol. 1, p. 151; *De multiplicatione specierum,* Bridges ed., Vol. 2, p. 453; *Communia naturalium,* Steele ed., fasc. 3, pp. 220, 224.

42. Bk. 1, lect. 42, Larcher tr., pp. 148–149; see also Bk. 2, lect. 20, pp. 235–240.

43. *Ibid.,* p. 149.

44. 87b39–88a3.

45. *Ibid.,* p. 149.

46. See Bk. 2, lect. 1, pp. 168–169.

47. Bk. 1, lect. 23, p. 75; see also Thomas Aquinas, *Exposition of Aristotle's Treatise On the Heavens,* tr. F. R. Larcher and P. H. Conway, 2 vols., Columbus: Ohio Dominican College, 1963 [pro manuscripto], Bk. 2, lect. 12, Vol. 2, p. 53.

48. *Ibid.,* p. 76.

49. Bk. 1, lect. 25, p. 81.

50. *Ibid.*

51. Bk. 2, lects. 6 & 7, pp. 184–192.

52. Bk. 2, lects. 8 & 9, pp. 192–201.

53. *Ibid.,* lect. 9, p. 200.

54. *Ibid.*

55. Other places where Aquinas discusses the difference between intrinsic and extrinsic final causes include *In XII Metaphysicorum,* lect. 12, no. 2627; *Contra Gentiles,* Bk. 4, cap. 97; *Summa theologiae,* Part I, q. 6, a. 3.

56. See *supra,* p. 46.

57. 390a3–5; cited *supra,* p. 18.

58. *Liber II Posteriorum,* tr. 2, cap. 11, Borgnet ed., Vol. 2, p. 192.

59. See Aquinas's *Commentary on Aristotle's Physics,* Bk. 2, lect. 3, nn. 160–161, Blackwell tr., pp. 78–79.

60. Aquinas, *Commentary on the Posterior Analytics of Aristotle,* Bk. 1, lect. 25, Larcher tr., pp. 80–81.

61. *Summa theologiae,* Part I, q. 1, a. 2 ad 2.

62. *Exposition of Aristotle's Treatise On the Heavens,* Bk. 2, lect. 28, Larcher and Conway tr., Vol. 2, pp. 113–115.

63. *Ibid.,* p. 113.

64. *Ibid.*

65. *Ibid.,* p. 114

66. *Ibid.*

67. *Ibid*.

68. *Ibid*. Copernicus's arguments are given in Bk. 1, 2, Great Books ed., Chicago: Britannica, 1952, Vol. 16, pp. 511–512.

69. *Ibid*., p. 115. For representative calculations and their relationship to modern measurements, see Aristotle, *On the Heavens*, Loeb Classical Library, tr. W. K. C. Guthrie, Cambridge, Mass.: Harvard U. P., 1939, p. 254, fn. a.

70. Lect. 27–28, pp. 110–113.

71. Lect. 27, p. 110.

72. Lect. 28, p. 113, translation adapted. I have rendered the imperfect tense of *dicebant* as "used to claim," rather than simply as "claimed," to stress the fact that the earth's sphericity was commonly accepted in Aquinas's time, long before the voyage of Columbus.

73. *Ibid*.

74. Probably from his study of Simplicius's commentary on *De caelo*. For a summary of Simplicius's teaching, see Duhem, *To Save the Phenomena*, pp. 22–24.

75. Six times in various works written between his commentary on the *Sentences* (*c.* 1256) and that on the *Metaphysics* (*c.* 1272), and five times in the commentary on the *De caelo* (1272–1273); for details, see Thomas Litt, *Les Corps célestes dans l'univers de saint Thomas d'Aquin*, Louvain: Publications Universitaires, 1963, pp. 322–366.

76. Bk. 2, lect. 11, Larcher and Conway tr., p. 47.

77. *Ibid*., lect. 17, pp. 74–77.

78. *Ibid*., pp. 74–75.

79. *Summa theologiae*, Part I, q. 32, a. 1 ad 2.

80. For details on Bernard of Verdun, see the article by Claudia Kren in the *Dictionary of Scientific Biography*, Vol. 2, pp. 23–24; for his teaching on eccentrics and epicycles, see Duhem, *To Save the Phenomena*, pp. 37–38, 40.

81. *Exposition of . . . On the Heavens*, Bk. 2, lect. 4, p. 18.

82. *Ibid*., lect. 17, p. 75.

83. *Ibid*., lect. 7, p. 31.

84. See Duhem, *To Save the Phenomena*, pp. 25–45; also Crombie, *Robert Grosseteste . . . ,* p. 97.

85. Translation by Giorgio de Santillana, *The Crime of Galileo*, Chicago: Chicago U. P., 1955, p. 99.

86. The best study is Erhard Schlund, "Peter Peregrinus von Maricourt, sein Leben und seine Schriften (Ein Beitrag zur Roger Baco-Forschung)" *Archivum Franciscanum Historicum* 4 (1911), pp. 436–455, pp. 633–643; 5 (1912), pp. 22–40. See also Bruno Rizzi, "Il magnetismo dalle origini e l'epistola *De magnete* de Pietro Peregrino," *Physis*, 11 (1969), pp. 502–519.

87. *Ibid*., 4 (1911), p. 447

88. *Opus tertium*, cap. 13. tr. by Bridges, *The Opus Majus . . . ,* Vol. 1, pp. xxv–xxvi.

89. I have used the Latin text edited by G. Hellmann, *Rara Magnetica 1269–1599*, Neudrucke von Schriften und Karten über Meteorologie und Erdmagnetismus 10 (1898), pp. (1)–(12), based on the *editio princeps* of Augsburg 1558; see Schlund, 5 (1912), pp. 36, 40.

90. Pars I, cap. 4, Hellmann ed., pp. (2)–(3).

91. *Ibid.*, p. (3).

92. *Ibid.*

93. *Ibid.*, cap. 5, pp. (3)–(4).

94. *Ibid.*, cap. 6, p. (4).

95. *Ibid.* Noteworthy is the juxtaposition of the word "experiment" with scammony attracting red bile; this suggests a knowledge on Peter's part of Galen or Avicenna, if not of Robert Grosseteste or Albert the Great.

96. *Ibid.*, cap. 7, pp. (4)–(5)

97. *Ibid.*, cap. 8, p. (5)

98. *Ibid.*, cap. 9, pp. (5)–(7)

99. Thus Peter's letter was known to William Gilbert when he wrote his classical treatise *De magnete* in 1600; see Bridges, *The Opus Maius* . . . , Vol. 1, p. 116, fn. 1; Vol. 2, p. 203, fn. 1. Gilbert also knew of, and praised, Aquinas's more philosophical treatment of the magnet in his letter *De occultis operibus naturae*, ed. J. B. McAllister, Washington, D.C.: Catholic University Press, 1939; see pp. 136–140, 187.

100. *Ibid.*, cap. 1, p. (1); the same preoccupation is evident, of course, in Aquinas's letter.

101. *Ibid.*, cap. 4, p. (2).

102. *Ibid.*, cap. 5, p. (3).

103. *Ibid.*, cap. 8, p. (5).

104. *Ibid.*, cap. 9, p. (5).

105. *Ibid.*, cap. 10, p. (7).

106. *Ibid.*

107. *Ibid.*, p. (8).

108. See Schlund, 4 (1911), p. 445.

109. The classical study is Engelbert Krebs, *Meister Dietrich, sein Leben, seine Werke, seine Wissenschaft*, Beiträge zur Geschichte der Philosophie des Mittelalters, Vol. 5, Parts 5–6, Münster: 1906.

110. Joseph Würschmidt, *Dietrich von Freiberg: Über den Regenbogen und die durch Strahlen erzeugten Eindrücke*, Beiträge zur Geschichte der Philosophie des Mittelalters, Vol. 12, Parts 5–6, Münster: 1914.

111. For a summary of Theodoric's life and works, see my *The Scientific Methodology of Theodoric of Freiberg*, Studia Friburgensia, N.S. no. 26, Fribourg: University Press, 1959, pp. 10–20.

112. *Robert Grosseteste* . . . , pp. 237–239.

113. This contrary to the statements of Crombie, Gilson, and

others that Theodoric subscribed to "light metaphysics"; see my *Scientific Methodology . . .* , p. 86.

114. *Scientific Methodology . . .* , *passim;* see also C. B. Boyer, *The Rainbow: From Myth to Mathematics,* New York: Thomas Yoseloff, 1959, pp. 110–124.

115. *Scientific Methodology . . .* , p. 177.

116. *Ibid.,* p. 178.

117. *Ibid.*

118. *Ibid.,* pp. 186–188.

119. *Ibid.,* p. 187 and references cited in fns. 1, 2, and 4.

120. Boyer, *The Rainbow,* p. 120.

121. *Ibid.,* pp. 119–122.

122. For excellent photographs of the MS drawings depicting these paths, see Crombie, *Robert Grosseteste . . .* , insert between pp. 256–257.

123. The demonstrations actually employed by Theodoric are given in schematic form in *Scientific Methodology,* pp. 215–218 for the primary rainbow, and pp. 223–225 for the secondary.

124. *De iride,* Würschmidt ed., p. 66

125. *Ibid.,* p. 34. I have edited the text of *De coloribus* in an appendix to *Scientific Methodology . . .* , pp. 364–376; for an analysis of the teaching it contains, see pp. 163–173.

126. *Scientific Methodology . . .* , pp. 183–205, 227–237, esp. p. 231.

127. *De iride,* pp. 34, 93, 94, 99, 102, 103, 106, 151, 183, 188, 193.

128. *Ibid.,* pp. 94, 95, 99.

129. For a fuller exposition, see *Scientific Methodology . . .* , pp. 237–245.

130. See *Scientific Methodology . . .* , pp. 168–169, 171. For Roger Bacon's teaching, see *Opus maius,* Part VI, cap. 8, Bridges ed., Vol. 2, pp. 190–191. Bacon was willing to admit, however, that the colors projected by the sun's rays passing through a prism have a "natural cause" and that this results in their being really present.

131. See Leff, *Paris and Oxford Universities . . .* , pp. 187–188, 222–240, 271–272, 290–294.

132. For further details, see Dijksterhuis, *The Mechanization . . .* , pp. 160–163.

133. A sympathetic presentation and evaluation of Duhem's thought is to be found in the introductory essay by S. L. Jaki in *To Save the Phenomena,* pp. ix–xxvi, esp. p. xix.

134. See the article by E. A. Moody on Buridan in the *Dictionary of Scientific Biography,* Vol. 2, pp. 603–608.

135. *Quaestiones super octo phisicorum libros,* Paris: Denis Roce, 1509, Lib. 2, q. 13, fol. 39ra–40vb. This is not to deny that there was a nominalist cast to Buridan's thought; for a discussion of this and Buridan's relation to Ockham, Robert Holkot, and Gre-

gory of Rimini, see T. K. Scott, "John Buridan on the Objects of Demonstrative Science," *Speculum*, 40 (1965), pp. 654–673. Scott bases his analysis on a manuscript of Buridan's *Quaestiones libri Ethicorum* (BN Ms Lat 16128), where in Question 6 of Book VI he takes up the interesting question "Whether everything knowable is eternal?" Buridan raises most of the problems relating to universal and scientific knowledge of contingent things that we have seen discussed by Grosseteste, Albert, and Aquinas, and offers a distinctive solution to them. He does not think, for example, that knowledge of a phenomenon like thunder is necessary only "in its causes," as some have proposed, because if a person knows that thunder is a sound in the clouds, then he has knowledge of thunder itself and not merely of what causes thunder. Again Buridan does not hold that universal knowledge has to be instantiated at every moment by some existing particular case, since it is proper to speak of knowledge of things even if no one of them exists, as in the case of knowing about thunder on fair days. Buridan has more sympathy, according to Scott, with the view that the objects of scientific knowledge are actually disguised hypothetical propositions even though stated in categorical form. "Thus the proposition 'Thunder is a sound in the clouds' is true even if there is no thunder at the time the proposition is written or uttered. For this is merely a disguised rendering of the proposition 'If there is thunder, then it is a sound in the clouds,' which is true quite independently of current weather conditions." — p. 668 Buridan's fuller solution depends on a theory of "natural supposition" that derives from the *Summulae* of Peter of Spain; this was generally absent from the Ockhamist tradition, and Scott expresses the opinion that more may be discovered about it "in further study of the logic of the realist tradition." — p. 670.

136. Buridan's philosophy of science is contained in his *Quaestiones in Metaphysicen* [sic] *Aristotelis*, Paris: Jodocus Badio, 1518, Lib. 2, qq. 1 and 2, fol. 8ra–10ra. For a summary of this, as well as the censure of Nicholas of Autrecourt, see E. A. Moody, *art. cit.*, *Dictionary of Scientific Biography*, Vol. 2, pp. 604–605.

137. *Quaestiones . . . physicorum*, Lib. 3, qq. 2, 6, 7, and 8.

138. *Ibid.*, q. 2, fol. 42rb.

139. *Ibid.*, q. 6, fol. 48vab.

140. *Ibid.*, q. 7, fol. 50va.

141. *Die Vorläufer Galileis . . .* , p. 22: "Bewegung ist ein ontologisch nicht weiter analysierbares, real gegebenes und empirisch feststellbares Moment am mobile, das von diesem und vom Ort verschieden ist, das aber nicht näher erklärt werden kann und auch nicht näher erklärt zu werden braucht."

142. This occurs in Franciscus's *Reportatio* on Bk. 4 of Peter Lombard's *Sentences*. The Latin text is given in A. Maier, *Zwei Grundprobleme der scholastischen Naturphilosophie*, Rome: Edizioni di Storia e Letteratura, 1951, pp. 166–180; partial Eng. tr. in

Clagett, *The Science of Mechanics* . . . , pp. 526–530, with commentary, pp. 530–531.

143. Maier, p. 173; Clagett, p. 529.

144. Maier, p. 177; Clagett, p. 530.

145. *Quaestiones . . . physicorum*, Lib. 8, q. 12, fol. 120rb–121rb; Latin text in Maier, *Zwei Grundprobleme* . . . , pp. 207–214; Eng. tr. Clagett, *The Science of Mechanics* . . . , pp. 532–538, with commentary, pp. 538–540.

146. *Quaestiones in Metaphysicen*, Lib. 12, q. 9, fol. 73ra.

147. *Quaestiones super libros quattuor de caelo et mundo*, ed. E. A. Moody, Cambridge, Mass.: Mediaeval Academy of America, 1942, Lib. 2, q. 12, pp. 176–181; Eng. tr. Clagett, *Science of Mechanics* . . . , pp. 557–562, with commentary, pp. 562–564.

148. Also in ways similar to those later used by Zabarella and Galileo; see C. B. Schmitt, "Experience and Experiment: A Comparison of Zabarella's View with Galileo's in *De motu*," *Studies in the Renaissance* 16 (1969), pp. 80–138, esp. p. 88 and fn. 19, p. 114 and fn. 84.

149. *Quaestiones super . . . de caelo*, Lib. 2, q. 12, Moody ed., p. 180; see Clagett, *Science of Mechanics*, p. 524 and fn. 39.

150. *Quaestiones . . . physicorum*, Lib. 8, q. 12, Clagett tr., pp. 533–534, 538.

151. *Science of Mechanics*, pp. 563–564.

152. *Quaestiones . . . physicorum*, Lib. 8, q. 12, Maier ed., p. 211; Clagett tr., p. 535.

153. *To Save the Phenomena*, p. 60.

154. Grant, "Hypotheses . . . ," p. 605.

155. *Quaestiones . . . physicorum*, Lib. 7, q. 8, fol. 108rb.

156. *Ibid.*, fol. 108va.

157. Grant, "Hypotheses . . . ," p. 605.

158. *Quaestiones in Metaphysicen*, Lib. 2, q. 1, fol. 9ra.

159. *Quaestiones super . . . de caelo*, Lib. 2, q. 22, Moody ed., p. 227; Clagett tr., p. 594.

160. *Ibid.*, Moody, pp. 228–229; Clagett, p. 595.

161. *Ibid.*

162. *Ibid.*, Moody, p. 229; Clagett, p. 595.

163. *Ibid.*

164. Granted, of course, that he saw reality, in this case, more through the eyes of faith than through reason.

165. *Dictionary of Scientific Biography*, Vol. 1, p. 94.

166. *Acutissime quaestiones super libros de physica auscultatione*, Venice: Octavianus Scotus, 1516.

167. *Ibid.*, fol. 36vb–37va.

168. *Ibid.*, fol. 37va–38ra.

169. *Ibid.*, fol. 37va.

170. See Crombie, *Medieval and Early Modern Science*, Vol. 2, pp. 73–74.

171. *Quaestiones . . . in libros de caelo*, Lib. 2, q. 14, text

and Eng. tr. in Clagett, *Science of Mechanics*, pp. 565–569; see also pp. 554–555.

172. *Dictionary of Scientific Biography*, Vol. 2, p. 607.

173. *Le Livre du ciel et du monde*, ed. A. D. Menut and A. J. Denomy, Madison: University of Wisconsin Press, 1968, Bk. 2, ch. 2, p. 289.

174. See Clagett, *Science of Mechanics*, pp. 552–553, 570–571, 635, 639, and 681.

175. *De proportionibus proportionum* and *Ad pauca respicientes*, ed. Edward Grant, Madison: University of Wisconsin Press, 1966.

176. I.e., between ratios of the type $(2/1)$ $^{1/2}$ and those of the type $(2/1)$ $^{1/}\sqrt{2}$.

177. See Grant, *De proportionibus proportionum*, pp. 61–65, 83, 304–305.

178. *Le Livre du ciel*, Bk. 1, ch. 24, pp. 167–179.

179. *Questiones de spera*, cited by Clagett, pre-print of article on Oresme to appear in the *Dictionary of Scientific Biography*, New York: American Council of Learned Societies, 1967, p. 8. Here Oresme seems to have had a better knowledge of the canons of demonstration than Galileo would have a century and a half later.

180. *Ibid.*, pp. 7–8.

181. *Le Livre du ciel*, Bk. 1, ch. 18, pp. 141–145.

182. This has been edited, translated, and commented on by Marshall Clagett, *Nicole Oresme and the Medieval Geometry of Qualities and Motions*, Madison: University of Wisconsin Press, 1968, pp. 157–517.

183. *Ibid.*, p. 15.

184. *Ibid.*, p. 297.

185. For a summary of this work, see John Murdoch's "*Mathesis in Philosophiam . . . ,*" and also his '*Rationes mathematice*': *Un Aspect du rapport des mathématiques et de la philosophie au moyen âge*, Paris: Université de Paris, 1967.

CHAPTER FOUR

1. See Leff, *Paris and Oxford Universities . . . ,* pp. 222–240.

2. Notably John Herman Randall, Jr., *The School of Padua and the Emergence of Modern Science*, Padua: Editrice Antenore, 1961. For a full bibliography, see Charles B. Schmitt, *A Critical Survey and Bibliography of Studies on Renaissance Aristotelianism 1958–1969*, Padua: Editrice Antenore, 1971; also helpful is Antonino Poppi, *Introduzione all'Aristotelismo Padovano*, Padua: Editrice Antenore, 1970.

3. See the article on Pietro by Loris Premuda in the *Dictionary of Scientific Biography*, Vol. 1, pp. 4–5.

4. These orders were discussed also by Albert the Great and

Thomas Aquinas; on the latter, see texts cited by Neal W. Gilbert, *Renaissance Concepts of Method*, New York: Columbia University Press, 1960, p. 29 and fns. 40–41.

5. *Conciliator differentiarum* . . . , ed. Venice 1496, Diff. 3, prop. 1, cited and translated by J. H. Randall, *The School of Padua* . . . , pp. 28–29.

6. Cited and translated from the *Galieni principis medicorum Microtegni cum commento Hali*, n.d., Randall, pp. 31–32.

7. Randall, p. 31.

8. Cited and translated from the Venice 1492 ed., comment on text 2, Randall, pp. 39–40. In the *Proemium* to his commentary on Aristotle's *Physics*, Averroës speaks of three modes of demonstration, namely, *demonstratio signi, demonstratio causae,* and *demonstratio simpliciter*. The first has been commonly understood as demonstration *quia*, the second as *propter quid*, and the third as combining both *quia* and *propter quid;* on this division, see John of St. Thomas, *The Material Logic of John of St. Thomas: Basic Treatises*, translated by Y. R. Simon, J. J. Glanville, and G. D. Hollenhorst. Chicago: University Press, 1955, pp. 495–500.

9. For a brief biography, see "Veneto, Paolo" by F. Roth in the *New Catholic Encyclopedia*, Vol. 14, p. 597.

10. *In Libros Posteriorum* . . . , Venice: Bonetus Locatellus, 1491, no foliation.

11. *Ibid.*, Lib. 1 comment on text [66] *In quibus autem media non convertuntur . . .* [78b11].

12. *Ibid.*, comment on chap. 13, text [65] *Item sic lunam demonstrant quod per incrementa circularis sit . . .* [78b4]

13. *Ibid.*

14. *Ibid.*, comment on chap. 30 at text *Sed eius quod est a fortuna non est scientia . . .* [87b18].

15. *Summa philosophiae naturalis*, Venice: Bonetus Locatellus, 1503, Lib.1, cap. 9, fol. 5rb, Randall tr., p. 40.

16. *Ibid.*, fol. 5va.

17. *Expositio super octo libros Physicorum Aristotelis*, Venice: Gregorius de Gregoriis, 1499, no foliation.

18. *Ibid.*, Lib. 1, comment on text 2, Randall tr., p. 41, adapted.

19. *Ibid.*, Lib. 3, comment on text 18, dub. 2.

20. *Ibid.*

21. *Ibid.*, Lib. 7, comment on text 2.

22. *Ibid.*, Lib. 3, comment on text 18, dub. 2, ad finem.

23. Cited and translated from the *Super Tegni Galeni*, ed. Padua 1475, comment on text 1, Randall, pp. 35–36.

24. See Randall, *The School of Padua* . . . , p. 22.

25. Cited and translated from the *Expositio super libros Tegni Galieni*, ed. Venice 1498, comment on text 1, Randall, pp. 37–38.

26. *The School of Padua* . . . , p. 38.

27. For a full account of the life and teachings of Gaetano see Silvestro da Valsanzibio, *Vita et Dottrina di Gaetano di Thiene,* Verona: Scuola Tipografica Madonna di Castelmonte, 1948.

28. *Regule* [*Hentisberi*] *cum sophismatibus. Declaratio Gaetani supra easdem* . . . Venice: Bonetus Locatellus, 1494.

29. *Regule* [Pars 6: Motus localis], text and translation in Clagett, *Science of Mechanics* . . . , pp. 235–236, 238–239.

30. *Declaratio Gaetani* . . . , fol. 38rb–va. Portions of the Latin text are given in my "Mechanics from Bradwardine to Galileo," *Journal of the History of Ideas* 32 (1971), p. 23 fns. 32–36.

31. See *Declaratio Gaetani* . . . , fol. 26ra, 28vb; also Valsanzibio, *Vita e Dottrina* . . . , pp. 198–202.

32. See his *Recollecte super octo libros Physicorum Aristotelis,* Venice: Octavianus Scotus, 1496, Lib. 1, q. 5.

33. For details, see Marshall Clagett, *Giovanni Marliani and Late Medieval Physics,* New York: Columbia University Press, 1941, p. 17, p. 139, and *passim.*

34. Clagett, *Giovanni Marliani* . . . , p. 140; Randall, *The School of Padua* . . . , pp. 22–23, also fn. 1 on p. 23.

35. Clagett, *Giovanni Marliani* . . . , pp. 139–140.

36. Hubert Élie, "Quelques maîtres de l'université de Paris vers l'an 1500," *Archives d'histoire doctrinale et littéraire du moyen âge,* 18 (1950–1951), pp. 193–243, esp. pp. 205–212.

37. In addition to Élie see A. Renaudet, *Préréforme et humanisme à Paris pendant les premières guerres d'Italie,* Paris: Bibliothèque de l'Institut Français de Florence, 1916, Première Série, Tome VI; and R. G. Villoslada, *La universidad de Paris durante los estudios de Francisco de Vitoria, O.P.,* 1507–1522, Rome: Analecta Gregoriana, 1938, Vol. XIV.

38. *Octo libri physicorum cum naturali philosophia atque metaphysica Johannis Maioris Hadingtonani theologi Parisiensis,* Paris: Johannes Parvus, 1526, no foliation.

39. *Ibid.,* Lib. 3, qq. 2–3.

40. For some brief textual excerpts, see my "The Concept of Motion in the Sixteenth Century," *Proceedings of the American Catholic Philosophical Association,* 41 (1967), pp. 189–190, fns. 29–32.

41. *Ibid.,* Lib. 7, q. [8].

42. *Questiones super octo libros phisicorum Aristotelis necnon super libros de celo et mundo,* Lyons: n.p., 1512.

43. *Ibid.,* Lib. 3, q. 1, fol. 54ra; Latin text in "Concept of Motion . . . ," p. 190, fn. 34.

44. *Ibid.,* fol. 61ra.

45. *Ibid.,* fol. 63rb–va; Latin text in "Concept of Motion . . . ," p. 191, fn. 36.

46. *Ibid.,* fol. 63va–115va.

47. On Coronel see my article in the *Dictionary of Scientific*

Biography, Vol. 3, pp. 420–421. I have used the edition of *Physice perscrutationes* published in Lyons by Simon Vincentius (n.d., but probably 1512); the first edition appeared in Paris in 1511.

48. See Duhem, *To Save the Phenomena . . .* , pp. 58–60.

49. *Physice perscrutationes,* Lib. 1, pars 1, fol. 1rb–3va.

50. *Ibid.,* fol. 2rb; Duhem tr., pp. 59–60.

51. *Ibid.,* Lib. 3, pars 1, fol. 52rb–6ova.

52. *Ibid.,* fol. 6ova; Latin text in my "Concept of Motion . . . ," p. 191, fn. 38.

53. *Ibid.,* Lib. 3, pars 4, fol. 84rb; see my "Concept of Motion . . . ," p. 192, fn. 40.

54. *Expositio magistri Joannis de Celaya Valentini in octo libros phisicorum Aristotelis, cum questionibus . . . secundum triplicem viam beati Thome, realium, et nominalium,* Paris: Hemundus le Feure, 1517.

55. See my "Concept of Motion . . . ," pp. 192–193.

56. Citations are from the Salamanca 1555 ed., although the first complete edition appeared at Salamanca in 1551, and an incomplete edition, lacking portions of Bk. 7 and all of Bk. 8, was published there *c.* 1545. For details relating to the work's preparation, see Vicente Beltrán de Heredia, *Domingo de Soto: Estudio biográfico documentado,* Salamanca: Biblioteca de Teologos Españoles, 1960, esp. p. 23.

57. *Super octo libros . . . , ed. cit.,* fol. 49ra–52ra.

58. *Ibid.,* fol. 50ra–b; some textual excerpts are given in my "Concept of Motion . . . ," pp. 193–194 and fns. 48–52.

59. *Ibid.,* Lib. 8, q. 3, fol. 99va–102rb.

60. *Ibid.,* Lib. 7, qq. 2–3, fol. 90ra–94rb.

61. *Ibid.,* q. 4, fol. 94rb–95vb.

62. Pierre Duhem, *Études sur Léonard de Vinci,* 3 vols., Paris: Hermann, 1906–1913, Vol. 3 (1913), pp. 263–583, esp. pp. 555–562.

63. *Super octo libros . . . ,* Lib. 7, q. 3, fol. 92vb.

64. For a full discussion of these systems of classification and their relationship to Soto, see my "The Enigma of Domingo de Soto: *Uniformiter difformis* and Falling Bodies in Late Medieval Physics," *Isis* 59 (1968), pp. 384–401.

65. *Ibid.,* pp. 387–396.

66. *Super octo libros . . . ,* Lib. 7, q. 3, fol. 92vb.

67. *Ibid.;* Latin text, "The Enigma . . . ," p. 400, fn. 60.

68. *Ibid.,* fol. 93vb; Latin text, "The Enigma . . . ," p. 400, fn. 61.

69. *Ibid.,* fol. 94ra; Latin text, "The Enigma . . . ," p. 400, fn. 62.

70. See C. B. Schmitt, "Experimental Evidence for and against a Void: The Sixteenth-Century Arguments," *Isis* 58 (1967), pp. 352–366.

71. For a fuller elaboration of Soto's position with respect to

those who preceded and came after him, see my "The 'Calcula-tores' in Early Sixteenth-Century Physics," *The British Journal for the History of Science* 4 (1969), pp. 221–232.

72. For a brief treatment see the article on Nifo by Michele Schiavone in the *Enciclopedia Filosofica*, 2d ed., 6 vols., Florence: G. C. Sansoni, 1967, Vol. 4, cols. 1032–1033. Nifo has often been characterized as an Averroist, but Edward P. Mahoney has recently shown that Nifo very early gave up Averroës as the true guide to Aristotle (mainly because Averroës used bad translations and could not read Greek), and turned his attention in great part to the Greek commentators, although he continued to maintain an interest in Aquinas. See Mahoney's "Agostino Nifo's Early Views on Immortality," *Journal of the History of Philosophy* 8 (1970), pp. 451–460; for other contributions by the same author, see C. B. Schmitt, *A Critical Survey* . . . , p. 165.

73. *Expositio super octo libros de physico auditu,* Venice: Oc-tavianus Scotus, 1508, Lib. 1, comm. text 4, fol. 7vb. On Averroës's three kinds of demonstration, see note 8 *supra.*

74. *Ibid.;* Randall tr., p. 42, adapted.

75. *Ibid.,* fol. 7vb–8ra; Randall tr., pp. 42–43, adapted.

76. See Randall, pp. 43–45.

77. Cited and translated from text 4 of the *Recognitio,* Ran-dall, pp. 44–46.

78. *In Aristotelis libros Posteriorum Analyticorum subtilissima commentaria,* Venice: Hieronymus Scotus, 1565, Lib. 1, fol. 11rb–va.

79. *Ibid.,* fol. 11va.

80. Cited and translated from text 4 of the *Recognitio,* Ran-dall, pp. 45–46.

81. *In . . . libros Posteriorum,* Lib. 1, fol. 11rb.

82. *Ibid.,* fol. 11va.

83. *Ibid.,* fol. 16ra.

84. Randall, *The School of Padua* . . . , p. 46. Not all the work on *regressus* during this period, however, was done explicitly in the context of discussions on methods of scientific discovery. See Antonino Poppi, "Pietro Pomponazzi tra averroismo e galenismo sul problema del 'regressus'," *Rivista Critica di Storia della Filoso-fia,* 24 (1969), pp. 243–266, which contains, in an appendix, a Latin edition of Pomponazzi's *Quaestio de regressu* (pp. 256–266).

85. Extracted from his commentary on the *Posterior Analytics,* Bk. 1, lect. 4, Larcher ed., p. 15.

86. *Expositio super octo libros* . . . , Lib. 2, fol. 66ra.

87. *Ibid.,* fol. 66rb.

88. *Ibid.,* Lib. 3, fol. 71vb.

89. See Randall, *The School of Padua* . . . , pp. 47–49; also Schmitt, *A Critical Survey* . . . , pp. 59–69.

90. For a brief biography see the article on Zabarella by John Glanville in the *New Catholic Encyclopedia,* Vol. 14, p. 1101;

fuller details are contained in a series of contributions by W. F. Edwards which are listed by C. B. Schmitt, *A Critical Survey* . . . , p. 153, and discussed *passim*.

91. *De methodis*, Lib. 1, cap. 2, in *Opera omnia quae ad perfectam Logicae cognitionem acquirendam spectare censetur utilissima*, 17th ed., Venice: Jacobus Sarzina, 1617, p. 71ab.

92. *Ibid.*, Lib. 3, cap. 3, p. 118b.

93. *Ibid.*, cap. 17, p. 139a.

94. *Ibid.*; Randall tr., p. 51.

95. *Ibid.*, cap. 18, p. 139b; Randall tr., p. 52.

96. *Ibid.*, p. 140a; Randall tr., p. 52.

97. *Ibid.*, cap. 19, p. 140b; Randall tr., p. 53.

98. *Ibid.*, p. 141b; Randall tr., p. 54, adapted.

99. *De regressu*, cap. 4, p. 251b; Randall tr., p. 56.

100. *Ibid.*, p. 252b; Randall tr., pp. 57–58.

101. *Ibid.*, p. 252b; Randall tr., p. 58, adapted.

102. Cf. Randall, *The School of Padua* . . . , p. 60.

103. See, however, the entries beginning "mathematica . . ." in the Index of the 1617 ed., fol. c2 preceding the pagination.

104. "Experience and Experiment: A Comparison of Zabarella's View with Galileo's in *De Motu*," *Studies in the Renaissance* 16 (1969), pp. 80–138.

105. *Ibid.*, p. 124.

106. *Ibid.*, p. 126.

107. *Ibid.*, p. 127.

108. Hieronymus Borrius Arretinus, *De motu gravium et levium*, Florence: Georgius Marescottus, 1576.

109. *Ibid.*, Pars 3, cap. 7, pp. 214–215.

110. *Ibid.*, p. 215. The Latin text reads as follows: ". . . cumque inter nos disceptationes semper augerentur, nullusque earumdem exitus inveniretur, ad experientiam omnium rerum magistram, perindeac ad sacram ancoram confugimus, et duobus adinventis, tum ligni, tum plumbi frustulis aequalis, ut ex aspectu coniicere licebat, ponderis, neque enim nos ad lancem illa expendere necessarium esse duximus, sed periculo quod facturi fueramus, sat esse rati fuimus, si ad oculum expenderentur, eadem ergo duo adinventa aequalis ponderis frustula, ex altiore nostrarum aedium fenestra pari impulsu, eodemque tempore proiiciemus: plumbum segnius descenderet, super lignum enim, quod prius in terram ceciderat, omnes quotquot ibi, rei exitum expectabamus, illud praeceps ruere vidimus: idque non semel, sed saepenumero eodem successu tentavimus. Cuius rei experimento ducti omnes in eamdem nobiscum pedibus iuêre sententiam. Ergo tum rationi, tum experimento, tum auctoritati consentaneum est, aeris in suo proprio, naturalique loco, nonnullam esse gravitatem, quo fit, ut lignum in quo plus est aeris, quam in plumbo, aequalis ponderis per aerem medium velocius descendat . . ."

111. See Schmitt, "Experience and Experiment . . . ," pp.

118–119 and especially fn. 95; also T. B. Settle, "Galileo's Use of Experiment as a Tool of Investigation," *Galileo Man of Science,* ed. Ernan McMullin, New York: Basic Books, 1967, pp. 324–326.

112. See Stillman Drake and I. E. Drabkin, *Mechanics in Sixteenth-Century Italy.* Selections from Tartaglia, Benedetti, Guido Ubaldo, and Galileo. Madison: University of Wisconsin Press, 1969, particularly the prefaces to Ubaldo's *Mechaniche,* pp. 241–258.

113. Duhem, *To Save the Phenomena,* p. 45.

114. Translated by Duhem, p. 47, from Achillini's *Omnia Opera,* Venice: Hieronymus Scotus, 1545, fol. 29b. This work was printed at Bologna in 1494 and a revised edition was issued in 1498; it appears in many subsequent editions of Achillini's *Opera.*

115. *Ibid.,* fol. 35b; Duhem tr., p. 48.

116. Thus Duhem states: "The quoted passages [from Nifo] are not a mere repetition of Averroës' teachings. They also bear the imprint, and recognizably so, of the thought of Thomas Aquinas, some of whose statements are reproduced word for word in Nifo's exposition." — *To Save the Phenomena,* p. 49.

117. Translated by Duhem, p. 48, from Nifo's *Aristotelis De Caelo et Mundo libri quatuor . . . ,* Venice: Hieronymus Scotus, 1549, Lib. 2, fol. 82 v.

118. *Ibid.*

119. See Duhem, *To Save the Phenomena,* pp. 49–52.

120. *Theorice nove planetarum Georgii Purbachii . . . ac in eas . . . Francisci Capuani de Manfredonia . . . expositio,* Venice: Simon Bevilacqua, 1495.

121. *Ibid.;* Duhem tr., p. 53.

122. Sylvester de Prierio, *In novas Georgii Purbachii theoricas planetarum commentaria,* Milan: 1514; also Paris: 1515.

123. *Ibid.;* Duhem tr., p. 53.

124. *To Save the Phenomena,* p. 56.

125. In his article on Copernicus in the *Dictionary of Scientific Biography,* Vol. 3, p. 410; see also Duhem, *To Save the Phenomena,* pp. 61–74. On Copernicus's studies at Padua, see Bruno Nardi, "Copernico studente a Padova," *Mélanges offerts à Étienne Gilson,* Toronto: Pontifical Institute of Mediaeval Studies, 1959, pp. 437–446.

126. See Max Jammer, *Concepts of Force,* Cambridge, Mass.: Harvard University Press, 1957, pp. 71–93.

CHAPTER FIVE

1. Galileo Galilei, *Dialogues Concerning the Two New Sciences,* tr. H. Crew and A. De Salvio. New York: Dover Publications, 1951, p. 165.

2. *Ibid.*

3. *Ibid.,* p. 166.

4. *Ibid.*, pp. 166–167.

5 Isaac Newton, *Mathematical Principles of Natural Philosophy,* tr. A. Motte, rev. and ed. by F. Cajori. Berkeley: University of California Press, 1934, p. 546.

6. *Ibid.*, p. 547, substituting "feign" for "frame" in the translation of "hypotheses non fingo."

7. For background on Gilbert, see Edgar Zilsel, "The Origins of William Gilbert's Methodology," *Journal of the History of Ideas* 2 (1941), pp. 1–32.

8. William Gilbert, *On the Loadstone and Magnetic Bodies,* Bk. 1, ch. 14. Chicago: Great Books of the Western World, 1952. Vol. 28, p. 20.

9. *Ibid.*, ch. 17, p. 24.

10. *Ibid.*

11. *Ibid.*, Bk. 2, ch. 2, p. 30.

12. *Ibid.*, ch. 4, pp. 36–37.

13. *Ibid.*

14. *Ibid.*, ch. 16, p. 47.

15. *Ibid.*, Bk. 3 ch. 2, p. 62.

16. *Ibid.*, Bk. 4, ch. 1, p. 77.

17. *Ibid.*

18. *Ibid.*

19. *Ibid.*, p. 78.

20. *Ibid.*, ch. 7, p. 82.

21. *Ibid.*, Bk. 5, ch. 1, p. 93.

22. *Ibid.*, p. 92.

23. *Ibid.*, title to ch. 11, p. 102.

24. *Ibid.*, ch. 11, p. 102.

25. *Ibid.*, ch. 12, p. 102.

26. *Ibid.*, title to ch. 12. The full title reads: "The magnetic force is animate, or imitates a soul; in many respects it surpasses the human soul while that is united to an organic body."

27. *Ibid.*, Bk. 6, ch. 3, p. 107.

28. *Ibid.*, ch. 4, p. 111: "the steadfastness of the poles," for example, "is controlled by the primary soul."

29. *Ibid.*

30. *Ibid.*, p. 112.

31. *Ibid.*, ch. 6, p. 116.

32. *Ibid.*

33. *Ibid.*, p. 117.

34. *Ibid.*, ch. 9, p. 121.

35. For details of Kepler's life and work, see Max Caspar, *Kepler: 1571–1630,* tr. C. D. Hellman, New York: Abelard-Schuman, 1959.

36. Written on October 13, 1597; see Carola Baumgardt, *Johannes Kepler: Life and Letters.* New York: Philosophical Library, 1941, p. 41.

37. *Epitome of Copernican Astronomy,* Bk. 4, tr. C. G. Wallis.

Chicago: Great Books of the Western World, 1952, Vol. 16, p. 850.

38. Kepler wished, of course, to go beyond Copernicus, "to ascribe to the earth on physical, or, if one prefers, metaphysical grounds the motion of the sun, as Copernicus does on mathematical grounds." Cited by Caspar, *Kepler*, p. 47.

39. This debt is acknowledged in Kepler's *Astronomia nova* (1609), whose full title translates as "A New Astronomy Based on Causality, or a Celestial Physics Derived from the Investigations of the Motions of the Star Mars, Founded on the Observations of the Noble Tycho Brahe."

40. *Epitome*, Bk. 4, p. 908.

41. *Ibid.*, p. 852.

42. Letter to Baron von Herberstein, May 15, 1596, tr. C. Baumgart, p. 33.

43. For a study of archetypal influence in Kepler's theories, see C. G. Jung and W. Pauli, *The Interpretation of Nature and the Psyche*, New York: Pantheon, 1955, p. 151 and ff.

44. *Joannis Kepleri Astronomi Opera Omnia*, ed. C. Frisch, 8 vols., Frankfurt am Main: 1858–1871, Vol. 1, p. 122.

45. *Timaeus*, 33B.

46. *Ibid.*, 34B.

47. *Epitome*, Bk. 4, title, p. 845.

48. E. A. Burtt, *Metaphysical Foundations . . .* , pp. 63–67.

49. *Ibid.*, p. 64.

50. Kepler, *Gesammelte Werke*, ed. W. von Dyck and Max Caspar, 18 vols., Munich: 1937–1951, Vol. 16, p. 196; tr. Max Jammer, *Concepts of Force*, pp. 84–85.

51. *Ibid.*, Vol. 3, p. 25; Jammer tr., p. 85.

52. *Ibid.*, p. 24.

53. Bk. 6, ch. 3, Great Books Ed., p. 109.

54. *Gesammelte Werke*, Vol. 3, p. 25; Jammer tr., p. 86.

55. *Ibid.*, p. 241; Jammer tr., p. 87.

56. E.g., in the *Tertius interveniens* (1610); see *Gesammelte Werke*, Vol. 4, p. 192; also Curtis Wilson, "From Kepler's Laws, So Called, to Universal Gravitation: Empirical Factors," *Archive for History of Exact Sciences*, 6 (1970), pp. 89–170.

57. See his letter to Maestlin of March 5, 1605, *Gesammelte Werke*, Vol. 15, p. 172, where he makes the statement that the sun's influence (*virtus*) "non est attractoria sed promotoria."

58. *Opera omnia*, ed. C. Frisch, Vol. 1, p. 176; Jammer tr., p. 90.

59. *Epitome*, Bk. 4, pp. 914–915.

60. *Ibid.*, pp. 916–917.

61. *Ibid.*, p. 917.

62. *Ibid.*, p. 931.

63. *Ibid.*, p. 899.

64. *Ibid.*

65. *Ibid.*, p. 902.

66. *Concepts of Force*, p. 92. Caspar comes to a similar conclusion: "It is Kepler's greatest service that he substituted a dynamic system for the formal schemes of the earlier astronomers, the law of nature for mathematical rule, and causal explanation for the mathematical description of nature." — *Kepler*, p. 135; see also p. 129.

67. For a summary treatment, which inclines, however, to an overly positivist interpretation of Galileo's thought, see Ludovico Geymonat, *Galileo Galilei*. A Biography and Inquiry into His Philosophy of Science, tr. Stillman Drake. New York: McGraw-Hill, 1965. For a non-positivist account, see W. R. J. Shea, "Galileo's Claim to Fame: The Proof That the Earth Moves From the Evidence of the Tides," *British Journal for the History of Science* 5 (1970), pp. 111–127.

68. *Le Opere di Galileo Galilei*, Edizione Nazionale, ed. Antonio Favaro, 20 vols., Florence: Tipografia di G. Barbera, 1890–1909, Vol. 1, pp. 15–177.

69. The *Quaestiones* on logic have yet to be edited, but an indication of their contents is provided by Favaro, *Le Opere*, Vol. 9, pp. 279–282, 291–292; one question is entitled "An detur regressus demonstrativus," fol. 31r–31v of MS Galileo 27, Biblioteca Nazionale, Florence.

70. For details, see my article on Buonamici, *Dictionary of Scientific Biography*, Vol. 2, pp. 590–591.

71. "Galileo and the Thomists," to appear in *St. Thomas Aquinas Commemorative Studies* (1274–1974), Pontifical Institute of Mediaeval Studies, Toronto, Canada. Upon receipt of a preliminary version of this paper, Professor A. C. Crombie of Oxford University has written me that his recent work on Galileo parallels my own, and that he will shortly publish, with the collaboration of Adriano Carugo and under the auspices of the Domus Galileiana in Pisa, a volume to be entitled *Galileo's Natural Philosophy,* which documents in detail the sources of Galileo's early writings and those of his mature thought.

72. This work is translated and analyzed by I. E. Drabkin in *Galileo Galilei. On Motion and On Mechanics*, I. E. Drabkin and S. Drake, trs. & eds., Madison: University of Wisconsin Press, 1960.

73. *Le Opere*, Vol. 1, p. 307; Drabkin tr., p. 76.

74. See C. B. Schmitt, "Experience and Experiment . . . ," p. 126; W. R. J. Shea, "The Significance of Experiments in the Writings of the Young Galileo," *Revue de l'Université d'Ottawa*, 41 (1971), pp. 198–199.

75. *Le Opere*, Vol. 1, p. 303; Drabkin tr., p. 71.

76. *Ibid.*, p. 300; Drabkin tr., p. 67.

77. *Ibid.*, p. 302; Drabkin tr., p. 70.

78. See Schmitt, "Experience and Experiment . . . ," p. 113,

and P. P. Wiener, "The Tradition Behind Galileo's Methodology," *Osiris* 1 (1936), p. 733. Galileo's gripe against contemporary Aristotelians is that their students "never know anything by its causes, but merely have opinions based on belief, that is, because this is what Aristotle said." — *Le Opere*, Vol. 1, p. 285; Drabkin tr., p. 50. Presumably he is correcting this defect by supplying knowledge through causes and not merely opinion based on belief.

79. For some elements of comparison between Galileo's and Zabarella's *De motu*, see Schmitt, "Experience and Experiment . . . ," pp. 107–108. Drabkin notes similarities between Galileo's and Buonamici's treatises in the footnotes to his translation on pp. 49, 55, 78, and 79; see also his bibliography on pp. 10–11.

80. *Le Opere*, Vol. 1, p. 260; Drabkin tr., p. 23.

81. *Ibid.*, p. 276; Drabkin tr., p. 41.

82. See J. A. Weisheipl, "Motion in a Void: Averroës and Aquinas," to appear in *St. Thomas Aquinas Commemorative Studies* (1274–1974), Pontifical Institute of Mediaeval Studies, Toronto, Canada.

83. *Le Opere*, Vol. 1, p. 314; Drabkin tr., p. 84.

84. *Ibid.*, p. 315; tr. p. 85.

85. *Ibid.*, pp. 315–316; tr. p. 85.

86. *Ibid.*, p. 319; tr. p. 89.

87. *Ibid.*, p. 322; tr. p. 93; see also p. 333 (tr. p. 105): "Now other writers, too, have tried to assign a cause for this remarkable effect [i.e., that "lighter bodies will fall more swiftly than heavier ones at the beginning of their motion"]. But since they did not master the problem, we shall in the next chapter refute their explanation and shall try to set forth the true cause."

88. *Le Opere*, Vol. 1, p. 337; tr. p. 110.

89. *Ibid.*, tr. p. 111.

90. On Galileo's use of *metodo risolutivo* and *metodo compositivo*, see Schmitt, "Experience and Experiment . . . ," pp. 128–129, 114, 85 fn. 11.

91. *Il Saggiatore*, in *Le Opere*, Vol. 6, p. 232.

92. On this point, see Burtt, *Metaphysical Foundations . . .* , pp. 74–83; also Geymonat, *Galileo Galilei*, pp. 31–34, 104–110, 169–174, 194–195.

93. *Dialogue Concerning the Two Chief World Systems*, tr. S. Drake, Berkeley: University of California Press, 1962, p. 406. In this connection, Galileo seems to have been more sympathetic to Kepler's methodology. Upon the receipt of the latter's *Mysterium cosmographicum* as a gift of the author, Galileo thanked Kepler in a letter of August 4, 1597, indicating his general agreement with the Copernican opinion, and stating that through it he had discovered "the causes of many natural effects" (cited by Geymonat, *Galileo*, pp. 23–24).

94. For a basic bibliography relating to the dispute, see Schmitt, "Experience and Experiment . . . ," p. 106, fn. 62.

95. This attitude runs all through the *De motu;* for example, see *Le Opere,* Vol. 1, pp. 273, 296, 300–301, 377; Drabkin tr., pp. 37–38, 40, 68.

96. *Ibid.,* p. 263; tr. p. 27. From these and similar statements Schmitt, "Experience and Experiment . . . ," p. 111, concludes that "for Galileo science deals with the 'causes of effects' and these cannot be reached through *experientia* — at least not through *experientia* alone."

97. Alexandre Koyré has drawn attention to this in his "Galileo's Treatise *De Motu Gravium:* The Use and Abuse of Imaginary Experiment," *Metaphysics and Measurement,* pp. 44–88.

98. *Two Chief World Systems,* Second Day, pp. 141–145.

99. *Two New Sciences,* p. 276.

100. *Two Chief World Systems,* pp. 234–235.

101. Geymonat and Drake seem to incline toward this view, possibly in reaction to the stress placed on Galileo's Platonism by Koyré and others; see *Galileo Galilei,* pp. 33, 133–135, 177–186; also T. R. Girill, "Galileo and Platonic Methodology," *Journal of the History of Ideas* 31 (1970), pp. 501–520.

102. Stillman Drake, *Galileo Studies:* Personality, Tradition, and Revolution, Ann Arbor: University of Michigan Press, 1970, p. 277. Descartes' criticism of Galileo, in fact, was that "without having considered the first causes of nature, he has only looked for the reasons for certain particular effects, and thus he has built without foundation" (Letter to Mersenne, October 11, 1638, in *Oeuvres,* ed. C. Adam and P. Tannery, 12 vols., Paris: 1897–1913, Vol. 2, p. 380). This, it would seem, is more a criticism of the lack of ultimacy in Galileo's explanations than in their causal character as such.

103. *Two New Sciences,* pp. 1–152. Similarly, in *The Assayer* (*Il Saggiatore*), Galileo discourses on the proposition "motion is the cause of heat" and on how one is to go about discovering the "true cause" of thermal effects; see the excerpts translated by Stillman Drake, *Discoveries and Opinions of Galileo,* Garden City: Doubleday, 1957, pp. 231–280, esp. pp. 272–279.

104. *Two New Sciences,* p. 276.

105. *Ibid.*

106. For a study of Galileo's atomism, see W. R. Shea, "Galileo's Atomic Hypothesis," *Ambix* 17 (1970), pp. 13–27.

107. This is not to deny that the concept of causal explanation undergoes a change in Galileo's successive expositions, particularly in his continuing study of local motion. At the outset he is concerned with efficient causes in the Aristotelian sense, while in the *Discorsi* he tends toward an explanation in the mathematical or formal sense, similar to that we have already seen in Kepler, and preparing for the idea of an objective law of nature that controls phenomena from within. On the latter notion, see Geymonat, *Galileo Galilei,* p. 179, where he cites Ernst Cassirer in

its support. For a more extended treatment of Galileo's rationalist tendencies despite his strong commitment to realism, see Maurice Clavelin, *La Philosophie naturelle de Galilée,* Paris: Librairie Armand Colin, 1968, pp. 456–459; also pp. 428–434. Clavelin is of the opinion that the analysis of falling motion in the *Discorsi* represents the first conjunction of the dynamic and kinematic points of view, "or, in traditional terminology, the point of view *quoad causas* and that *quoad effectus* (ou, comme le disait la tradition, entre le point de vue *quoad causas* et le point de vue *quoad effectus*)" — p. 359.

108. *Galileo Galilei,* p. 183.

109. *Ibid.*

110. D'Arcy Powers, *William Harvey,* London: 1897, pp. 26–27; cited by H. A. Ratner, "William Harvey, M.D.: Modern or Ancient Scientist?" *The Thomist* 24 (1961), p. 175.

111. Contained in the standard English edition of the collected works, *The Works of William Harvey, M.D.,* translated from the Latin by Robert Willis, London: Sydenham Society, 1847; reprinted New York: Johnson Reprint Co., 1965. For the extent of Harvey's allegiance to Aristotle, see Walter Pagel, "William Harvey Revisited," *History of Science,* 8 (1969), pp. 1–31, 9 (1970), pp. 1–41; Jerry Stannard, "Aristotelian Influences and References in Harvey's *De motu locali animalium*," Richard Tursman, ed., *Studies in Philosophy and the History of Science:* Essays in honor of Max Fisch. Lawrence, Kans.: Coronado Press, 1970, pp. 122–131; G. K. Plochmann, "William Harvey and His Methods," *Studies in the Renaissance,* 10 (1963), pp. 192–210; and J. S. Wilkie, "Harvey's Immediate Debt to Aristotle and Galen," *History of Science,* 4 (1965), pp. 103–124.

112. *Works,* p. 162.

113. *Ibid.*

114. *Ibid.*

115. *On the Generation of Animals,* Bk. 3, ch. 10; see *supra,* Chap. 1, note 30.

116. *Works,* p. 162.

117. *Ibid.,* p. 163.

118. *Ibid.,* pp. 166–167; Fabricius, of course, was Harvey's teacher at Padua. For fuller details on the *De generatione* and its relation to the *De motu cordis,* see Charles Webster, "Harvey's *De generatione:* Its Origin and Relevance to the Theory of Circulation," *British Journal for the History of Science* 3 (1967), pp. 262–274.

119. Title of chap. 14, *Works,* p. 68.

120. These notes have been preserved and are reproduced in facsimile and transcript in William Harvey, *Prelectiones anatomiae universalis,* London: J. & A. Churchill, 1886; significant excerpts are in Charles Singer, *A History of Biology,* rev. ed., New York: Henry Schuman, 1950, pp. 108–114.

121. *Works*, p. 45.
122. *Ibid.*, pp. 45–46.
123. *Ibid.*, p. 51.
124. *Ibid.*, pp. 51–52.
125. *Ibid.*, p. 51.
126. Chaps. 10–13, *ibid.*, pp. 52–67.
127. *Ibid.*, p. 68, italics added.
128. *Ibid.*, p. 71.
129. *Ibid.*, p. 86, reading "cause" for "purpose" in Willis's translation.
130. See Ratner, "William Harvey," pp. 178–181.
131. Thomas Fowler, cited by Ratner, p. 180.
132. *Works*, p. 114.
133. *Ibid.*, p. 98.
134. *Ibid.*, pp. 115–116.
135. *Ibid.*, p. 116.
136. *Ibid.*, p. 117.
137. *Ibid.* Earlier, in his treatise on animal generation, Harvey had insisted that "there is, in fact, no occasion for searching after spirits foreign to, or distinct from, the blood" (*ibid.*, p. 502), while in his original paper on the motion of the heart and blood he had maintained that "the blood and spirits constitute one body (like whey and butter in milk, or heat [and water] in hot water . . .)" — *ibid.*, p. 12.
138. *Ibid.*, pp. 122–123.
139. *Ibid.*, p. 123.
140. Bk. 1, chap. 31 (87b39–88a4) and chap. 34 (89b10–15).
141. *Ibid.*, p. 124; note, again, the terminology of the *Posterior Analytics:* "that which is in fact," or the answer to the question *an sit?*, and "the reason wherefore it is so," the answer to the question *propter quid?*
142. *Ibid.*, p. 133.
143. Harvey did not use any Latin term that could be translated by the word "pump," and yet this seems to have been his intention. See Chauncey D. Leake, preface to the fifth edition of his translation of William Harvey, M.D., *Anatomical Studies on the Motion of the Heart and Blood.* Springfield, Ill.: Charles C. Thomas, 1970, p. xxviii.
144. *Ibid.*, p. 137. It is in this sense that we can agree with Gweneth Whitteridge's observation that Harvey "does not fly back from his observations to first principles; he seeks only the immediate and efficient cause of the phenomenon." *William Harvey and the Circulation of the Blood,* New York: American Elsevier, 1971, p. xii. Less felicitous, in our opinion, is her conviction "that Harvey falls into the category of the great scientist who is not conscious of any philosophical method underlying his actions" (p. xi), a conviction that seems belied by the methodological statements we have documented in some detail.

145. *Ibid.;* for a detailed contrast of Harvey's teaching with that of Descartes, see J. A. Passmore, "William Harvey and the Philosophy of Science," *Australasian Journal of Philosophy,* 36 (1958), pp. 85–94.

146. *Physics,* Bk. 2, ch. 1, 192b21–23; most commentators held that this definition of nature was directly applicable to the soul, as the substantial form of an animated body.

147. For a brief but accurate study of Newton, see the article on him by R. S. Westfall in the *New Catholic Encyclopedia,* Vol. 10, pp. 424–428. Newton's contributions to mechanics are detailed by Westfall in his *Force in Newton's Physics. The Science of Dynamics in the Seventeenth Century,* New York: American Elsevier, 1971.

148. "The long passages quoted from Aristotle and his commentators [in Newton's Cambridge University Library notebook] are a reminder that, though Newton was an early convert to the new philosophy, he was first educated in the traditional philosophic disciplines." — F. E. Manuel, *A Portrait of Isaac Newton,* Cambridge, Mass.: Harvard University Press, 1968, p. 13.

149. John Herivel, *The Background to Newton's* PRINCIPIA. A Study of Newton's Dynamical Researches in the Years 1664–84, Oxford: Clarendon Press, 1965, p. 125, fn. 3.

150. *Ibid.,* p. 126.

151. R. S. Westfall, "The Foundations of Newton's Philosophy of Nature," *British Journal for the History of Science* 1 (1962), pp. 171–182.

152. Robert Hooke was one of the early critics of Newton's optical papers, and apparently Newton "resolved to publish no more in this field while Hooke lived." — E. N. da C. Andrade, *Sir Isaac Newton,* Garden City: Doubleday, 1958, p. 101; significantly, Hooke died in 1703.

153. Newton's early optical papers have been described and analyzed by T. S. Kuhn in I. B. Cohen, ed., *Isaac Newton's Papers and Letters on Natural Philosophy and Related Documents,* Cambridge, Mass.: Harvard University Press, 1958, pp. 27–45; the papers themselves are reprinted on pp. 47–238. See also R. S. Westfall, "The Development of Newton's Theory of Color," *Isis* 53 (1962), pp. 339–358.

154. *The Correspondence of Isaac Newton,* ed. H. W. Turnbull, Vol. 1, Cambridge: University Press, 1959, pp. 96–97.

155. Particularly Newton's reply to Anthony Lucas, *Philosophical Transactions,* No. 128, Sept. 25, 1676, pp. 702–704; Cohen ed., pp. 173–175.

156. A. I. Sabra, *Theories of Light From Descartes to Newton,* New York: American Elsevier Co., 1967, p. 248.

157. *Philosophical Transactions,* No. 80, Feb. 19, 1671/72, p. 3076; Cohen ed., p. 48.

158. For a description and analysis, see Sabra, *Theories of*

Light, pp. 231–250; also my *Scientific Methodology*, pp. 268–278. An evaluation that differs from my own is Torger Holtsmark, "Newton's *Experimentum crucis* Reconsidered," *American Journal of Physics* 38 (1970), pp. 1229–1235.

159. *Philosophical Transactions*, No. 80 (1671/72), p. 3079; Cohen ed., p. 51.

160. *Ibid.*

161. *Opticks*, Bk. 3, Part 1, 4th ed., London: 1730, reprinted New York: Dover Publications, 1952, pp. 404–405; also my *Scientific Methodology*, p. 283.

162. *Scientific Methodology*, p. 274.

163. *Philosophical Transactions*, No. 80 (1671/72), p. 3081; Cohen ed., p. 53.

164. *Ibid.*, pp. 3081–82; Cohen ed., pp. 53–54.

165. L. Rosenfeld, "La théorie des couleurs de Newton et ses adversaires," *Isis* 9 (1927), pp. 60–61.

166. *Philosophical Transactions*, No. 88, Nov. 18, 1672, p. 5087; Cohen ed., p. 119 (English modernized).

167. *Ibid.*, No. 97, Oct. 6, 1673, p. 6109; Cohen ed., p. 144.

168. *Ibid.*, No. 84, June 17, 1672, p. 4087; English translation in *Philosophical Transactions Abridged*, eds. Hutton, Shaw, and Pearson, London: 1809, Vol. 1, p. 726; Cohen ed., pp. 79, 86.

169. *Ibid.*, p. 732; Cohen ed., p. 92.

170. *Ibid.*, p. 739; Cohen ed., p. 105.

171. *Ibid.*, p. 740; Cohen ed., p. 106.

172. *Ibid.*, p. 743; Cohen ed., p. 109.

173. For details on Line and Lucas, see T. S. Kuhn, "Newton's Optical Papers," in Cohen, *Isaac Newton's Papers . . .*, p. 34, fn. 13; also Conor Reilly, "Francis Line, Peripatetic (1595–1675)," *Osiris* 14 (1962), pp. 221–253. Lucas succeeded Line as professor of physics at Liège and continued Line's controversy with Newton after Line's death.

174. *Philosophical Transactions*, No. 128, Sept. 25, 1676, pp. 702–704; Cohen ed., pp. 173–175 (English modernized).

175. Sabra, *Theories of Light*, pp. 243–250, 273–297; see also the review essay of Sabra's book by R. S. Westfall, "The Science of Optics in the Seventeenth Century," *History of Science*, 6 (1967), pp. 150–156, and the equally challenging essay by P. K. Feyerabend, "Classical Empiricism," in R. E. Butts and J. W. Davis, eds., *The Methodological Heritage of Newton*, Toronto: University of Toronto Press, 1970, pp. 150–170, esp. pp. 158–169.

176. *Philosophical Transactions*, No. 80 (1671/72), p. 3081; Cohen ed., p. 53. For a stimulating discussion of a related topic in Newton's *Opticks*, with similar philosophical implications, see R. S. Westfall, "Uneasily Fitful Reflections on Fits of Easy Transmission," *The* ANNUS MIRABILIS *of Sir Isaac Newton 1666–1966*, ed. Robert Palter, Cambridge, Mass.: The M.I.T. Press, 1970, pp. 88–104.

177. This topic has been examined with great care by Alexandre Koyré, *Newtonian Studies*, Cambridge, Mass.: Harvard University Press, 1965, pp. 53–200.

178. As revised by Florian Cajori, *Sir Isaac Newton's Mathematical Principles of Natural Philosophy and His System of the World*, Berkeley: University of California Press, 1934. A fuller study will have to await the appearance of the *variorum* edition of this work, now being prepared by I. Bernard Cohen. In the interim, see Cohen's "Newton's Second Law and the Concept of Force in the *Principia*," *The* ANNUS MIRABILIS . . . , ed. Robert Palter, pp. 143–185, and the many works referenced in the notes, pp. 171–176.

179. See R. E. Butts, "Whewell on Newton's Rules of Philosophizing," *Methodological Heritage*, p. 136.

180. These are given in their entirety, with the exception of the explanation of the third rule, for which only the first sentence is given.

181. Cajori ed., pp. 398–400. For the variant readings of the different editions, see Koyré, *Newtonian Studies*, pp. 261–272.

182. Thus Newton was not a Platonist or a Neopythagorean to the same degree as were Galileo or Kepler, although he was somewhat under the influence of Cambridge Platonists such as Henry More, to whose thought his teacher, Isaac Barrow, was also sympathetic. See Burtt, *Metaphysical Foundations* . . . , pp. 150–161; also E. W. Strong, "Barrow and Newton," *Journal of the History of Philosophy* 8 (1970), pp. 155–172.

183. Cajori ed., p. 399. There is, of course, a problem with Newton's transition from a gravitational principle within a single observed body to a "principle of *mutual* gravitation" [italics mine], as was pointed out by Roger Cotes when raising the "invisible hand" objection. For a discussion of this extremely interesting point, and the way in which Newton seemingly handled the difficulty in light of his own metaphysics, see Koyré, *Newtonian Studies*, pp. 273–282, esp. p. 274 and pp. 280–281.

184. *Ibid.*

185. *Ibid.*, p. 400. Two recent and illuminating studies of Newton's use of causal argument and induction are Dudley Shapere, "The Philosophical Significance of Newton's Science," *The* ANNUS MIRABILIS . . . , ed. Robert Palter, pp. 285–299, and Robert Palter, "Newton and the Inductive Method," *ibid.*, pp. 244–257.

186. See Gerd Buchdahl, "Gravity and Intelligibility: Newton to Kant," *Methodological Heritage*, pp. 74–102, esp. pp. 78–81. For a general account of Newton's physical and metaphysical commitments, see A. J. Snow, *Matter and Gravity in Newton's Physical Philosophy. A Study in the Natural Philosophy of Newton's Time*, Oxford: University Press, 1926.

187. Cajori ed., p. 546.

188. *Ibid.*, p. 547. Thus we disagree with the thesis of H. G. Van Leeuwen, *The Problem of Certainty in English Thought 1630–1690*, The Hague: Martinus Nijhoff, 1963, pp. 106–120, where he attempts to show that Newton gave up on the possibility of achieving any certitude in scientific knowledge. Newton's skepticism was with respect to ultimate explanations, and here we can agree with Van Leeuwen's conclusions, but Newton entertained no such skepticism with regard to his knowledge of proximate causes and explanations of the phenomena he had experimentally investigated.

189. *Ibid.*, pp. 546–547, substituting "feign" for "frame" as in note 6 *supra*. See Koyré, *Newtonian Studies*, pp. 35–36.

190. *Newtonian Studies*, pp. 115–184.

191. *Ibid.*, p. 182.

192. *Ibid.*, p. 118.

193. *Ibid.*, p. 127.

194. *Ibid.*, pp. 139–148. For background, see Marie Boas [Hall], "The Establishment of the Mechanical Philosophy," *Osiris* 10 (1952), pp. 412–541.

Bibliography

The following list contains only materials cited and used directly in the preparation of this study. Entries under "Literature" that are starred with an asterisk (°) are noteworthy for the additional bibliography they provide. In the listings under "Sources," works in Latin are given before English translations and, where applicable, the ordering follows that of the *corpus aristotelicum* rather than that of simple alphabetization.

SOURCES

Achillini, Alessandro, *Opera omnia in unum collecta* . . . , Venice: Hieronymus Scotus, 1545.

Albert the Great, *Opera omnia*, ed. Augustine Borgnet, 38 vols., Paris: Ludovicus Vivès, 1890–1899.

———, *Opera omnia*, ed. Bernhard Geyer, Cologne: Institutum Alberti Magni, 1951–.

Albert of Saxony, *Commentarius in Posteriora [analyticorum] Aristotelis*. Venice: Bonetus Locatellus, 1497.

———, *Acutissime quaestiones super libros de physica auscultatione*, Venice: Octavianus Scotus, 1516.

———, *Quaestiones subtilissime* . . . *in libros de caelo et mundo*, Venice: Bonetus Locatellus, 1492.

Aquinas, St. Thomas, *Opera omnia*. Rome: Commissio Leonina, 1882–.

———, *In libros posteriorum analyticorum expositio*, cura et studio R. M. Spiazzi, Rome-Turin: Marietti, 1955.

———, *In octo libros physicorum Aristotelis expositio*, cura et studio P. M. Maggiòlo, Rome-Turin: Marietti, 1954.

———, *In duodecim libros metaphysicorum Aristotelis expositio*, cura et studio R. M. Spiazzi, Rome-Turin: Marietti, 1950.

Aquinas, St. Thomas, *De occultis operibus naturae* and *De motu cordis*, in *Opuscula philosophica*, ed. R. M. Spiazzi, Rome-Turin: Marietti, 1954.

———, *Commentary on the Posterior Analytics of Aristotle*, English translation by F. R. Larcher, Albany: Magi Books, 1970.

———, *Commentary on Aristotle's Physics*, English translation by R. J. Blackwell *et al.*, New Haven: Yale University Press, 1963.

———, *Exposition of Aristotle's Treatise On the Heavens*, English translation by F. R. Larcher and P. H. Conway, 2 vols., Columbus: Ohio Dominican College, 1963 [pro manuscripto].

———, *De occultis operibus naturae*, English translation in J. B. McAllister, *The Letter of Saint Thomas Aquinas De Occultis Operibus Naturae ad Quemdam Militem Ultramontanum.* Philosophical Studies, Vol. 42. Washington, D.C.: The Catholic University of America Press, 1939.

Archimedes, *Measurement of a Circle*, ed. T. L. Heath, *The Works of Archimedes.* New York: Dover Publications, 1953.

Aristotle, *Opera*, ed. Immanuel Bekker, 5 vols., Berlin: Academia Regia Borussica, 1831–1870.

———, *Analytica Posteriora*, ed. L. Minio-Paluello & B. G. Dod, *Aristoteles Latinus*, Vol. 4, 1–4. Bruges-Paris: Desclée de Brouwer, 1968.

———, *The Works of Aristotle.* Translated into English under the Editorship of W. D. Ross, 12 vols., Oxford: University Press, 1908–1952:
Prior Analytics, tr. A. J. Jenkinson
Posterior Analytics, tr. G. R. G. Mure
Sophistic Refutations, tr. W. A. Pickard-Cambridge
Physics, tr. R. P. Hardie & R. K. Gaye
On the Heavens, tr. J. L. Stocks
On the Generation of Animals, tr. Arthur Platt

———, The Loeb Classical Library. Cambridge, Mass.: Harvard University Press, 1912–.
Posterior Analytics, tr. H. Tredennick & E. S. Forster
Physics, tr. P. H. Wicksteed & F. M. Cornford
On the Heavens, tr. W. K. C. Guthrie
Meteorologica, tr. H. D. P. Lee

———, *Aristotle's Physics.* Translated with Commentaries and Glossary by Hippocrates G. Apostle. Bloomington: Indiana University Press, 1969.

———, *Aristotle's Physics I, II*, tr. W. Charlton, New York: Oxford University Press, 1970.

———, *Aristotle's Physics*, tr. Richard Hope, Lincoln: University of Nebraska Press, 1961.

———, *Aristotle's Metaphysics*, tr. Richard Hope, New York: Columbia University Press, 1952.

Averroës, *Aristotelis opera cum Averrois commentariis* . . . , ed. Marcus Antonius Zimara. Venice: Apud Juntas, 1562–1574.

——, *Aristotelis physica cum commentario Averrois*, editio princeps, Padua: n.p., *c.* 1472–1475.

Bacon, Roger, *Opus majus. The 'Opus Majus' of Roger Bacon*, ed. J. H. Bridges, 2 vols., Oxford: Clarendon Press, 1897.

——, *The 'Opus Majus' of Roger Bacon*, English translation by R. B. Burke, 2 vols., Philadelphia: University of Pennsylvania Press, 1928.

Benedetti, Giovanni Battista, *Demonstratio proportionum motuum localium contra Aristotelem et omnes philosophos.* Venice: n.p., 1554.

——, *Diversarum speculationum mathematicarum et physicarum liber.* Turin: Haeres Bevilaquae, 1585.

——, *Resolutio omnium Euclidis problematum aliorumque ad hoc necessario inventorum* . . . , Venice: Bartholomaeus Caesanus, 1553.

Bernard of Verdun, *Tractatus super totam astrologiam,* ed. Polykarp Hartmann. Franziskanische Forschungen, 15. Werl (Westfalen): Dietrich Coelde Verlag, 1961.

Borro, Girolamo, *De motu gravium et levium.* Florence: Georgius Marescottus, 1576.

Bradwardine, Thomas, *Tractatus de proportionibus,* Edited and translated by H. Lamar Crosby, Jr., in *Thomas of Bradwardine: His "Tractatus de Proportionibus."* Its Significance for the Development of Mathematical Physics. Madison: University of Wisconsin Press, 1955.

Buonamici, Francesco, *De motu libri decem quibus generalia naturalis philosophiae principa summo studio collecta continentur necnon universae quaestiones ad libros De physico auditu, De caelo, De ortu et interitu pertinentes explicantur* . . . , Florence: Bartholomaeus Sermartellius, 1591.

Buridan, Jean, *Quaestiones super octo phisicorum libros.* Paris: Denis Roce, 1509.

——, *Quaestiones super libris quattuor de caelo et mundo,* ed. Ernest A. Moody, Cambridge, Mass.: Mediaeval Academy of America, 1942.

——, *Questiones in Metaphysicen Aristotelis.* Paris: Jodocus Badio, 1518.

Burley, Walter, *Super octo libros physicorum.* Venice: Bonetus Locatellus, 1491.

——, *De intensione et remissione formarum.* Venice: Octavianus Scotus, 1496.

Capuano of Manfredonia, Francesco, *Theorice nove planetarum Georgii Purbachii . . . ac in eas . . . Francisci Capuani de Manfredonia . . . expositio.* Venice: Simon Bevilacqua, 1495.

Celaya, Juan de, *Expositio magistri Joannis de Celaya Valentini in octo libros phisicorum Aristotelis, cum quaestionibus . . . se-*

cundum triplicem viam beati Thome, realium, et nominalium.
Paris: Hemundus le Feure, 1517.

Copernicus, Nicholas, *De revolutionibus orbium coelestium libri sex.* Nuremberg: Iohannes Petreius, 1543.

——, *On the Revolutions of the Heavenly Spheres.* Great Books of the Western World, Vol. 16. Chicago: Encyclopaedia Britannica, 1952.

Coronel, Luis Nuñez, *Physice perscrutationes.* Lyons: Simon Vincentius, n.d. [1512?].

Descartes, René, *Oeuvres,* ed. C. Adam & P. Tannery, 12 vols., Paris: Léopold Cerf, 1897–1913.

Dullaert of Ghent, Jean, *Questiones super octo libros phisicorum Aristotelis necnon super libros de celo et mundo,* Lyons: n.p., 1512.

Euclid, *Opera omnia,* ed. J. L. Heiberg. Leipzig: Teubner, 1895.

Gaetano da Thiene, *Regule* [*Hentisberi*] *cum sophismatibus. Declaratio Gaetani supra easdem* . . . Venice: Bonetus Locatellus, 1494.

——, *Recollecte super octo libros physicorum Aristotelis,* Venice: Octavianus Scotus, 1496.

Galileo Galilei, *Le Opere di Galileo Galilei,* Edizione Nazionale, ed. Antonio Favaro, 20 vols., Florence: Tipografia di G. Barbera, 1890–1909.

——, *Dialogue Concerning the Two Chief World Systems,* tr. Stillman Drake, Berkeley: University of California Press, 1962.

——, *Dialogues Concerning the Two New Sciences,* tr. H. Crew & A. De Salvio, New York: Dover Publications, 1951.

——, *On Motion* and *On Mechanics,* tr. and ed. I. E. Drabkin & Stillman Drake, Madison: University of Wisconsin Press, 1960.

Gilbert, William, *De magnete, magneticisque corporibus, et de magno magnete tellure* . . . , London: Petrus Short, 1600.

——, *On the Loadstone and Magnetic Bodies.* Great Books of the Western World, Vol. 28. Chicago: Encyclopaedia Britannica, 1952.

Giles of Rome, *In libros posteriorum Aristotelis expositio.* Venice: Bonetus Locatellus, 1488.

——, *In libros de physico auditu Aristotelis commentaria* . . . , Venice: Bonetus Locatellus, 1502.

Grosseteste, Robert, *Posteriorum* [*analyticorum*] *Lincolniensis* . . . , Venice: Octavianus Scotus, 1521.

——, *Commentarius in octo libros physicorum Aristotelis,* ed. Richard C. Dales, Boulder: University of Colorado Press, 1963.

——, *De artibus liberalibus, De cometis et causis ipsarum, De generatione sonorum, De iride, De lineis angulis et figuris seu de fractionibus et reflexionibus radiorum, De luce,* and *De sphaera,* ed. Ludwig Baur, *Die Philosophie des Robert Grosse-*

teste, Bischofs von Lincoln. Beiträge zur Geschichte der Philosophie des Mittelalters, Band XVIII, 4–6, Münster: 1917.
——, *De luce,* English translation by C. C. Riedl, *Robert Grosseteste on Light,* Milwaukee: Marquette University Press, 1942.

Harvey, William, *The Works of William Harvey, M.D.,* translated from the Latin by Robert Willis. London: Sydenham Society, 1847; reprinted New York: Johnson Reprint Co., 1965.
——, *Anatomical Studies on the Motion of the Heart and Blood,* translated and edited by Chauncey D. Leake, 5th ed., Springfield, Ill.: Charles C. Thomas, 1970.

Heytesbury, William, *Tractatus . . . de sensu composito et diviso. Regule eiusdem cum sophismatibus . . . ,* Venice: Bonetus Locatellus, 1494.

John of St. Thomas, *The Material Logic of John of St. Thomas: Basic Treatises,* tr. Y. R. Simon, J. J. Glanville, and G. D. Hollenhorst. Chicago: University of Chicago Press, 1955.

Kepler, Johannes, *Gesammelte Werke,* ed. W. von Dyck & Max Caspar, 18 vols, Munich: 1937–1951.
——, *Epitome of Copernican Astronomy,* tr. C. G. Wallis. Great Books of the Western World, Vol. 16. Chicago: Encyclopaedia Britannica, 1952.
——, *Letters.* Edited and translated in Carola Baumgardt, *Johannes Kepler: Life and Letters,* New York: Philosophical Library, 1941
——, *Opera omnia,* ed. C. Frisch, 8 vols., Frankfurt am Main: Heyder & Zimmer, 1858–1871.

Leibniz, Gottfried Wilhelm, *Letters,* in *The Leibniz-Clarke Correspondence,* together with extracts from Newton's *Principia* and *Opticks,* edited with introduction and notes by H. G. Alexander, New York: Philosophical Library, 1956.

Mair, Jean, *Octo libri physicorum cum naturali philosophia atque metaphysica Johannis Maioris Hadingtonani theologi Parisiensis,* Paris: Johannes Parvus, 1526.
——, *Propositum de infinito,* edited and translated into French by Hubert Élie, *Le Traité "De l'Infini" de Jean Mair.* Nouvelle édition avec traduction et annotations. Paris: J. Vrin, 1938.

Newton, Isaac, *Mathematical Principles of Natural Philosophy,* tr. A. Motte, rev. and ed. by F. Cajori. Berkeley: University of California Press, 1934.
——, *Opticks,* 4th ed., London: 1730; reprinted New York: Dover Publications, 1952.
——, *The Correspondence of Isaac Newton,* ed. H. W. Turnbull, Vol. 1, Cambridge: University Press, 1959.
——, *Isaac Newton's Papers and Letters on Natural Philosophy and Related Documents,* ed. I. B. Cohen, Cambridge, Mass.: Harvard University Press, 1958.
——, *Unpublished Scientific Papers of Isaac Newton.* A Selection

from the Portsmouth Collection in the University Library, Cambridge, chosen, edited and translated by A. R. Hall and M. B. Hall. Cambridge: University Press, 1962.

Nifo, Agostino, *In Aristotelis libros posteriorum analyticorum subtilissima commentaria*. Venice: Hieronymus Scotus, 1565.

——, *Expositio super octo libros de physico auditu*. Venice: Octavianus Scotus, 1508.

——, *Aristotelis de caelo et mundo libri quatuor*. Venice: Hieronymus Scotus, 1549.

Ockham, William of, *Summa totius logicae*, ed. Philotheus Boehner, St. Bonaventure, N.Y.: Franciscan Institute, 1951.

——, *Tractatus de successivis*, ed. Philotheus Boehner, St. Bonaventure, N.Y.: Franciscan Institute, 1944.

——, *Philosophical Writings: A Selection*, translated with an introduction by Philotheus Boehner, Edinburgh: Thomas Nelson, 1957; New York: Bobbs-Merrill, 1964.

Oresme, Nicole, *De proportionibus proportionum* and *Ad pauca respicientes*, ed. Edward Grant, with introductions, English translations, and critical notes. Madison: University of Wisconsin Press, 1966.

——, *Le Livre du ciel et du monde*, ed. A. D. Menut and A. J. Denomy, translation and introduction by A. D. Menut. Madison: University of Wisconsin Press, 1968.

——, *Tractatus de configurationibus qualitatum et motuum*, edited with an introduction, English translation, and commentary by Marshall Clagett, *Nicole Oresme and the Medieval Geometry of Qualities and Motions*, Madison: University of Wisconsin Press, 1968.

Paul of Venice, *In libros posteriorum [analyticorum Aristotelis]* . . . , Venice: Bonetus Locatellus, 1491.

——, *Expositio super octo libros physicorum Aristotelis*, Venice: Gregorius de Gregoriis, 1499.

——, *Summa philosophiae naturalis*. Venice: Bonetus Locatellus, 1503.

Peckham, John, *Perspectiva communis*, edited with an introduction, English translation, and critical notes by David C. Lindberg. Madison: University of Wisconsin Press, 1970.

Peter Peregrinus of Maricourt, *De magnete seu rota perpetui motus libellus*, editio princeps, Augsburg: 1558, ed. G. Hellmann, *Rara Magnetica 1269–1599*, Neudrucke von Schriften und Karten über Meteorologie und Erdmagnetismus 10 (1898), pp. (1)–(12).

Plato, *The Collected Dialogues of Plato Including the Letters*, edited by Edith Hamilton and Huntington Cairns, with introduction and prefatory notes. Bollingen Series LXXI. Princeton: University Press, 1961.

Purbach, Georg, *Theoricae novae planetarum* . . . , Wittenberg: Joannes Lufft, 1542.

Soto, Domingo de, *Super octo libros physicorum quaestiones*, Salamanca: Juan de Junta, 1545?; Salamanca: A. de Portonaris, 1551–1552; 2d ed., Salamanca: A. de Portonaris, 1555–1556.
Swineshead, Richard, *Calculationes noviter emendate atque revise* . . . , Venice: Heres Octaviani Scoti, 1520.
———, *De loco elementi* [Tractatus 11 calculationum], edited with English translation by M. A. Hoskin & A. G. Molland, "Swineshead on Falling Bodies: An Example of Fourteenth-Century Physics," *British Journal for the History of Science* 3 (1966), pp. 150–182.
Theodoric of Freiberg, *De iride et radialibus impressionibus*, ed. Joseph Würschmidt, *Dietrich von Freiberg: Über den Regenbogen und die durch Strahlen erzeugten Eindrücke*, Beiträge zur Geschichte der Philosophie des Mittelalters, Band XII, 5–6, Münster: 1914.
———, *De accidentibus, De coloribus, De elementis*, and *De miscibilibus in mixto*, ed. W. A. Wallace, *The Scientific Methodology of Theodoric of Freiberg*, Studia Friburgensia, N.S. no. 26, Fribourg: University Press, 1959.
Zabarella, Jacopo, *Opera omnia quae ad perfectam logicae cognitionem acquirendam spectare censetur utilissima*, 17th ed., Venice: Jacobus Sarzina, 1617.

LITERATURE

Andrade, E. N. da C., *Sir Isaac Newton*. Garden City: Doubleday, 1958.
Baumgardt, Carola, *Johannes Kepler: Life and Letters*. New York: Philosophical Library, 1941.
Beltrán de Heredia, Vicente, *Domingo de Soto*. Estudio biográfico documentado. Salamanca: Biblioteca de Teologos Españoles, Vol. 20, 1960. Madrid: Ediciones Cultura Hispánica, 1961.
Bohm, David, *Causality and Chance in Modern Physics*. New York: D. Van Nostrand, 1957.
Born, Max, *Natural Philosophy of Cause and Chance*, Oxford: University Press, 1949.
Boyer, Carl B., *The History of the Calculus and Its Conceptual Development*, New York: Dover Publications, 1959.
———, *The Rainbow: From Myth to Mathematics*, New York: Thomas Yoseloff, 1959.
* Brody, Baruch A., ed., *Readings in the Philosophy of Science*, Englewood Cliffs: Prentice-Hall, 1970.
Bromberger, Sylvain, "Why-Questions," *Readings in the Philosophy of Science*, B. A. Brody, ed., Englewood Cliffs: Prentice-Hall, 1970, pp. 66–87.
Buchdahl, Gerd, "Gravity and Intelligibility: Newton to Kant," *The Methodological Heritage of Newton*, R. E. Butts & J. W.

Davis, eds., Toronto: University of Toronto Press, 1970, pp. 74–102.

Buchdahl, Gerd, *Metaphysics and the Philosophy of Science.* The Classical Origins: Descartes to Kant. Cambridge, Mass.: M.I.T. Press, 1969.

Bunge, Mario, *Causality:* The Place of the Causal Principle in Modern Science, Cambridge, Mass.: Harvard University Press, 1959.

Burtt, E. A., *The Metaphysical Foundations of Modern Science,* 2d rev. ed., New York: Humanities Press, 1932.

Butts, R. E., "Whewell on Newton's Rules of Philosophizing," *The Methodological Heritage of Newton,* R. E. Butts & J. W. Davis, eds., Toronto: University of Toronto Press, 1970, pp. 132–149.

Butts, R. E., & Davis, J. W., eds., *The Methodological Heritage of Newton,* Toronto: University of Toronto Press, 1970.

° Callus, Daniel, ed., *Robert Grosseteste: Scholar and Bishop,* Oxford: University Press, 1955.

Caspar, Max, *Kepler: 1571–1630,* tr. C. D. Hellman, New York: Abelard-Schuman, 1959.

Clagett, Marshall, *Giovanni Marliani and Late Medieval Physics,* New York: Columbia University Press, 1941.

———, *Nicole Oresme and the Medieval Geometry of Qualities and Motions,* Madison: University of Wisconsin Press, 1968.

———, "Oresme, Nicole," pre-print of article to appear in the *Dictionary of Scientific Biography,* New York: American Council of Learned Societies, 1967.

° ———, *The Science of Mechanics in the Middle Ages.* Madison: University of Wisconsin Press, 1959.

Clavelin, Maurice, *La Philosophie naturelle de Galilée.* Essai sur les origines et la formation de la mécanique classique. Paris: Librairie Armand Colin, 1968.

Cohen, I. Bernard, ed., *Isaac Newton's Papers and Letters on Natural Philosophy and Related Documents,* Cambridge, Mass.: Harvard University Press, 1958.

———, "Newton's Second Law and the Concept of Force in the *Principia,*" *The* ANNUS MIRABILIS *of Sir Isaac Newton 1666–1966,* ed. Robert Palter, Cambridge, Mass.: The M.I.T. Press, 1970, pp. 143–185.

° Crombie, Alistair C., *Medieval and Early Modern Science,* 2 vols., New York: Doubleday Anchor, 1959 (revised edition of *Augustine to Galileo,* Cambridge, Mass.: Harvard University Press, 1953).

° ———, *Robert Grosseteste and the Origins of Experimental Science,* Oxford: University Press, 1953.

° Crombie, A. C. & North, J. D., "Bacon, Roger," *Dictionary of Scientific Biography,* Vol. 1, pp. 377–385.

Dales, Richard C., "Robert Grosseteste's *Commentarius in Octo Li-*

bros Physicorum Aristotelis," *Medievalia et Humanistica,* 11 (1957), pp. 10–33.

——, "Robert Grosseteste's Scientific Works," *Isis* 52 (1961), pp. 381–402.

Dijksterhuis, E. J., *The Mechanization of the World Picture,* tr. C. Dikshoorn, Oxford: University Press, 1961.

* Drake, Stillman, *Discoveries and Opinions of Galileo,* Garden City: Doubleday, 1957.

——, *Galileo Studies:* Personality, Tradition, and Revolution. Ann Arbor: University of Michigan Press, 1970.

* Drake, S., & Drabkin, I. E., *Mechanics in Sixteenth-Century Italy.* Selections from Tartaglia, Benedetti, Guido Ubaldo, and Galileo. Madison: University of Wisconsin Press, 1969.

Duhem, Pierre, *Études sur Léonard de Vinci.* Ceux qu'il a lus et ceux qui l'ont lu, 3 vols., Paris: Hermann, 1906–1913.

——, *To Save the Phenomena,* tr. E. Doland and C. Maschler, Chicago: University Press, 1969.

Eastwood, Bruce S., "Grosseteste's 'Quantitative' Law of Refraction: A Chapter in the History of Non-experimental Science," *Journal of the History of Ideas* 28 (1967), pp. 403–414.

——, "Medieval Empiricism: The Case of Grosseteste's Optics," *Speculum* 43 (1968), pp. 306–321.

——, "Robert Grosseteste's Theory of the Rainbow," *Archives internationales d'histoire des sciences* 78 (1966), pp. 313–332.

Élie, Hubert, "Quelques maîtres de l'université de Paris vers l'an 1500," *Archives d'histoire doctrinale et littéraire du moyen âge* 18 (1950–1951), pp. 193–243.

——, *Le Traité "De l'Infini" de Jean Mair.* Nouvelle édition avec traduction et annotations. Paris: J. Vrin, 1938.

Evans, Melbourne G., "Causality and Explanation in the Logic of Aristotle," *Philosophy and Phenomenological Research* 19 (1958–1959), pp. 466–485.

Feyerabend, Paul K., "Classical Empiricism," *The Methodological Heritage of Newton,* eds. R. E. Butts & J. W. Davis, Toronto: University of Toronto Press, 1970, pp. 150–170.

Geymonat, Ludovico, *Galileo Galilei:* A Biography and Inquiry into His Philosophy of Science, tr. Stillman Drake, foreword by Giorgio de Santillana. New York: McGraw-Hill Book Co., 1965.

Gilbert, Neal W., *Renaissance Concepts of Method,* New York: Columbia University Press, 1960.

Gillispie, Charles C., ed., *Dictionary of Scientific Biography,* New York: Charles Scribner's Sons, 1970–.

Gilson, Étienne, "Pourquoi S. Thomas a critiqué S. Augustin," *Archives d'histoire doctrinale et littéraire du moyen âge* 1 (1926), pp. 1–126.

Girill, T. R., "Galileo and Platonistic Methodology," *Journal of the History of Ideas* 31 (1970), pp. 501–520.

Glanville, John, "Zabarella, Jacopo," *The New Catholic Encyclopedia*, 15 vols., New York: McGraw-Hill Book Co., 1967, Vol. 14, p. 1101.

Grant, Edward, "Hypotheses in Late Medieval and Early Modern Science," *Daedalus*, Proceedings of the American Academy of Arts and Sciences, 91 (1962), pp. 599–616.

° ———, *Nicole Oresme. 'De proportionibus proportionum' and 'Ad pauca respicientes,'* edited with introductions, English translations, and critical notes. Madison: University of Wisconsin Press, 1966.

[Hall], Marie Boas, "The Establishment of the Mechanical Philosophy," *Osiris* 10 (1952), pp. 412–541.

Harré, Rom[ano], *Matter and Method*. New York: St. Martin's Press, 1964.

° ———, *The Principles of Scientific Thinking*. Chicago: University Press, 1970.

———, *Theories and Things*. London: Sheed & Ward, 1961.

Heisenberg, Werner, *The Physicist's Conception of Nature*, tr. A. J. Pomerans, London: Hutchinson, 1958.

Hempel, Carl G., & Oppenheim, Paul, "Studies in the Logic of Explanation," *Philosophy of Science* 15 (1948), pp. 135–175.

Herivel, John, *The Background to Newton's 'Principia.'* A Study of Newton's Dynamical Researches in the Years 1664–84. Oxford: Clarendon Press, 1965.

Hesse, Mary B., *Forces and Fields. The Concept of Action at a Distance in the History of Physics*. London: Thomas Nelson, 1961.

———, *Models and Analogies in Science*. Notre Dame: University Press, 1966.

Holtsmark, Torger, "Newton's *Experimentum crucis* Reconsidered," *American Journal of Physics* 38 (1970), pp. 1229–1235.

Hoskin, M. A., & Molland, A. G., "Swineshead on Falling Bodies: An Example of Fourteenth-Century Physics," *British Journal for the History of Science* 3 (1966), pp. 150–182.

Jaki, Stanley L., "Introductory Essay," in Pierre Duhem, *To Save the Phenomena*, tr. E. Doland & C. Maschler, Chicago: University Press, 1969, pp. ix–xxvi.

———, *The Relevance of Physics*, Chicago: University Press, 1966.

Jammer, Max, *Concepts of Force*, Cambridge: Mass.: Harvard University Press, 1957.

Jung, C. G., & Pauli, W., *The Interpretation of Nature and the Psyche*, New York: Pantheon Books, 1955.

Koyré, Alexandre, *Metaphysics and Measurement*. Essays in the Scientific Revolution. Cambridge, Mass.: Harvard University Press, 1968.

———, *Newtonian Studies*, Cambridge, Mass.: Harvard University Press, 1965.

——, Review of A. C. Crombie, *Robert Grosseteste* . . . , in *Diogenes*, 16 (1956), pp. 1–22.

Krebs, Engelbert, *Meister Dietrich, sein Leben, seine Werke, seine Wissenschaft*, Beiträge zur Geschichte der Philosophie des Mittelalters, Vol. 5, 5–6, Münster: 1906.

Kren, Claudia, "Bernard of Verdun," *Dictionary of Scientific Biography*, Vol. 2, pp. 23–24.

Kuhn, Thomas S., "Newton's Optical Papers," *Isaac Newton's Papers and Letters on Natural Philosophy and Related Documents*, ed. I. B. Cohen, Cambridge, Mass.: Harvard University Press, 1958, pp. 27–45.

——, *The Structure of Scientific Revolutions*, Chicago: University Press, 1962; 2d enlarged ed., 1970.

° Leff, Gordon, *Paris and Oxford. Universities in the Thirteenth and Fourteenth Centuries:* An Institutional and Intellectual History, New York: John Wiley & Sons, 1968.

Lenzen, Victor F., *Causality in Natural Science*, Springfield, Ill.: Charles C. Thomas, 1954.

Lindberg, David C., "The Cause of Refraction in Medieval Optics," *The British Journal for the History of Science* 4 (1968), pp. 23–38.

° ——, *John Pecham and the Science of Optics. Perspectiva communis*, edited with an introduction, English translation, and critical notes. Madison: University of Wisconsin Press, 1970.

——, "Lines of Influence in Thirteenth-Century Optics: Bacon, Witelo, and Pecham," *Speculum* 46 (1971), pp. 66–83.

——, "A Reconsideration of Roger Bacon's Theory of Pinhole Images," *Archive for History of Exact Sciences* 6 (1970), pp. 214–223.

——, "Roger Bacon's Theory of the Rainbow: Progress or Regress?" *Isis* 57 (1966), pp. 235–248.

——, "The Theory of Pinhole Images from Antiquity to the Thirteenth Century," *Archive for History of Exact Sciences* 5 (1968), pp. 154–176.

——, "The Theory of Pinhole Images in the Fourteenth Century," *Archive for History of Exact Sciences* 6 (1970), pp. 299–325.

Litt, Thomas, *Les Corps célestes dans l'univers de saint Thomas d'Aquin*, Louvain: Publications Universitaires, 1963.

Lorentz, H. A., et al., *The Principle of Relativity.* A Collection of Original Memoirs on the Special and General Theory of Relativity, tr. W. Perrett & G. B. Jeffery, London: Methuen, 1923; reprinted New York: Dover Publications, n.d.

Lynch, L. E., "The Doctrine of Divine Ideas and Illumination in Robert Grosseteste," *Mediaeval Studies* 3 (1941), pp. 161–173.

Mahoney, Edward P., "Agostino Nifo's Early Views on Immortality," *Journal of the History of Philosophy* 8 (1970), pp. 451–460.

Mahoney, Edward P., "A Note on Agostino Nifo," *Philological Quarterly* 50 (1971), pp. 125–132.

Maier, Anneliese, *Die Vorläufer Galileis im 14. Jahrhundert*, Rome: Edizioni di Storia e Letteratura, 1949.

———, *Zwei Grundprobleme der scholastischen Naturphilosophie*, Rome, Edizioni di Storia e Letteratura, 1951.

Manuel, Frank E., *A Portrait of Isaac Newton*. Cambridge, Mass.: Harvard University Press, 1968.

° McMullin, Ernan, ed., *Galileo Man of Science*, New York: Basic Books, 1967.

Merlan, Philip, "Bryson," *Dictionary of Scientific Biography*, Vol. 2, pp. 549–550.

Minio-Paluello, L., "Aristotle: Tradition and Influence," *Dictionary of Scientific Biography*, Vol. 1, pp. 267–281.

Minio-Paluello, L., & Dod, B. G., eds., *Aristoteles Latinus*, Vol. 4, 1–4, Bruges-Paris: Desclée de Brouwer, 1968.

Mittelstrass, Jürgen, *Die Rettung der Phänomene: Ursprung und Geschichte eines antiken Forschungsprinzips*, Berlin: Walter de Gruyter, 1962.

Moody, E. A., "Albert of Saxony," *Dictionary of Scientific Biography*, Vol. 1, pp. 93–95.

° ———, "Buridan, Jean," *Dictionary of Scientific Biography*, Vol. 2, pp. 603–608.

° Murdoch, John E., "Bradwardine, Thomas," *Dictionary of Scientific Biography*, Vol. 2, pp. 390–397.

———, "*Mathesis in philosophiam scholasticam introducta*: The Rise and Development of the Application of Mathematics in Fourteenth-Century Philosophy and Theology," *Arts libéraux et philosophie au moyen âge*. Actes du quatrième congrès international de philosophie médiévale, 1967. Montreal: Institut d'Études Médiévales, 1969, pp. 215–254.

———, '*Rationes mathematice*': *Un Aspect du rapport des mathématiques et de la philosophie au moyen âge*, Paris: Université de Paris, 1967.

° Murdoch, J. E., & Sylla, Edith, "Burley, Walter," *Dictionary of Scientific Biography*, Vol. 2, pp. 608–612.

Nagel, Ernest, *The Structure of Science*. Problems in the Logic of Scientific Explanation. New York: Harcourt, Brace and World, 1961.

Nardi, Bruno, "Copernico studente a Padova," *Mélanges offerts à Étienne Gilson*. Toronto: Pontifical Institute of Mediaeval Studies, 1959, pp. 437–446.

Nash, Leonard K., *The Nature of the Natural Sciences*. Boston: Little, Brown & Co., 1963.

Neurath, Otto, *et al.*, *Foundations of the Unity of Science*, 2 vols., Chicago: University Press, 1970.

Owen, G. E. L., "Aristotle: Method, Physics, and Cosmology," *Dictionary of Scientific Biography*, Vol. 1, pp. 250–258.

——, "*Tithenai ta phainomena*," *Aristote et les problèmes de méthode*. Communications présentées au Symposium Aristotelicum tenu à Louvain du 24 août au 1er septembre 1960. Louvain: Publications Universitaires, 1961, pp. 83–103.

° Pagel, Walter, "William Harvey Revisited," *History of Science* 8 (1969), pp. 1–31; 9 (1970), pp. 1–41.

Palter, Robert, ed., *The* ANNUS MIRABILIS *of Sir Isaac Newton 1666–1966*. Cambridge, Mass.: The M.I.T. Press, 1970.

——, "Newton and the Inductive Method," *The* ANNUS MIRABILIS *of Sir Isaac Newton 1666–1966*, ed. Robert Palter, Cambridge, Mass.: The M.I.T. Press, 1970, pp. 244–257

Passmore, John A., "William Harvey and the Philosophy of Science," *Australasian Journal of Philosophy* 36 (1958), pp. 85–94.

Plochmann, G. K., "William Harvey and His Methods," *Studies in the Renaissance* 10 (1963), pp. 192–210.

Pohle, William, "The Mathematical Foundations of Plato's Atomic Physics," *Isis* 62 (1971), pp. 36–46.

Popper, Karl R., *The Logic of Scientific Discovery*, New York: Basic Books, 1959.

° Poppi, Antonino, *Introduzione all'Aristotelismo Padovano*. Padua: Editrice Antenore, 1970.

——, "Pietro Pomponazzi tra averroismo e galenismo sul problema del 'regressus'," *Rivista Critica di Storia della Filosofia* 24 (1969), pp. 243–266.

Post, H. R., "Correspondence, Invariance, and Heuristics: In Praise of Conservative Induction," *Studies in History and Philosophy of Science* 2 (1971), pp. 213–255.

Premuda, Loris, "Abano, Pietro d'," *Dictionary of Scientific Biography*, Vol. 1, pp. 4–5.

Randall, John Herman, Jr., *The School of Padua and the Emergence of Modern Science*, Padua: Editrice Antenore, 1961.

Ratner, H. A., "William Harvey, M.D.: Modern or Ancient Scientist?" *The Thomist*, 24 (1961), pp. 175–208.

Reilly, Conor, "Francis Line, Peripatetic (1595–1675)," *Osiris* 14 (1962), pp. 221–253, 368.

——, *Francis Line S.J., An Exiled English Scientist 1595–1675*. Rome: Institutum Historicum S.I., 1969.

Renaudet, Augustin, *Préréforme et humanisme à Paris pendant les premières guerres d'Italie 1494–1517*, Paris: Bibliothèque de l'Institut Français de Florence, 1916, Première Série, Tome VI.

Rizzi, Bruno, "Il magnetismo dalle origini e l'epistola *De magnete* de Pietro Peregrino," *Physis*, 11 (1969), pp. 502–519.

° Rosen, Edward, "Copernicus," *Dictionary of Scientific Biography*, Vol. 3, pp. 401–411.

Rosenfeld, L., "La Théorie des couleurs de Newton et ses adversaires," *Isis* 9 (1927), pp. 44–65.

Roth, Francis G., "Veneto, Paolo," *New Catholic Encyclopedia*, Vol. 14, p. 597.

Sabra, A. I., *Theories of Light From Descartes to Newton*, New York: American Elsevier Co., 1967.

Sambursky, S., *The Physical World of the Greeks*, tr. M. Dagut, New York: Macmillan, 1956.

Santillana, Giorgio de, *The Crime of Galileo*, Chicago: University Press, 1955.

Schiavone, Michele, "Nifo, Agostino," *Enciclopedia Filosofica*, 2d ed., 6 vols., Florence: G. C. Sansoni, 1967, Vol. 4, cols. 1032–1033.

Schlund, Erhard, "Peter Peregrinus von Maricourt, sein Leben und seine Schriften (Ein Beitrag zur Roger Baco-Forschung)," *Archivum Franciscanum Historicum*, 4 (1911), pp. 436–455, 633–643; 5 (1912), pp. 22–40.

° Schmitt, Charles B., *A Critical Survey and Bibliography of Studies on Renaissance Aristotelianism 1958–1969*. Padua: Editrice Antenore, 1971.

° ———, "Experience and Experiment: A Comparison of Zabarella's View with Galileo's in *De motu*," *Studies in the Renaissance* 16 (1969), pp. 80–138.

———, "Experimental Evidence for and against a Void: The Sixteenth-Century Arguments," *Isis* 58 (1967), pp. 352–366.

Scott, Theodore K., "John Buridan on the Objects of Demonstrative Science," *Speculum* 40 (1965), pp. 654–673.

Scriven, Michael, "Explanations, Predictions, and Laws," *Readings in the Philosophy of Science*, ed. B. A. Brody, Englewood Cliffs: Prentice-Hall, 1970, pp. 88–104.

Settle, Thomas B., "Galileo's Use of Experiment as a Tool of Investigation," *Galileo Man of Science*, ed. Ernan McMullin, New York: Basic Books, 1967, pp. 315–337.

Shapere, Dudley, "The Philosophical Significance of Newton's Science," *The ANNUS MIRABILIS of Sir Isaac Newton 1666–1966*, ed. Robert Palter, Cambridge, Mass.: The M.I.T. Press, 1970, pp. 285–299.

Shapiro, Herman, *Motion, Time and Place According to William Ockham*, St. Bonaventure, N.Y.: Franciscan Institute, 1957.

Sharp, D. E., "The *De ortu scientiarum* of Robert Kilwardby (d. 1279)," *The New Scholasticism* 8 (1934), pp. 1–30.

Shea, William R., "Galileo's Atomic Hypothesis," *Ambix* 17 (1970), pp. 13–27.

———, "Galileo's Claim to Fame: The Proof That the Earth Moves From the Evidence of the Tides," *British Journal for the History of Science* 5 (1970), pp. 111–127.

———, "The Significance of Experiments in the Writings of the

Young Galileo," *Revue de l'Université d'Ottawa* 41 (1971), pp. 192–206.

Singer, Charles, *A History of Biology*, rev. ed., New York: Henry Schuman, 1950.

Snow, A. J., *Matter and Gravity in Newton's Physical Philosophy.* A Study in the Natural Philosophy of Newton's Time. Oxford: University Press, 1926.

Sprague, R. K., "The Four Causes: Aristotle's Exposition and Ours," *The Monist* 52 (1968), pp. 298–300.

Stannard, Jerry, "Aristotelian Influences and References in Harvey's *De motu locali animalium*," *Studies in Philosophy and the History of Science:* Essays in Honor of Max Fisch, ed. Richard Tursman, Lawrence, Kans.: Coronado Press, 1970, pp. 122–131.

Strong, Edward W., "Barrow and Newton," *Journal of the History of Philosophy* 8 (1970), pp. 155–172.

Thorndike, Lynn, *A History of Magic and Experimental Science*, 8 vols., New York: Columbia University Press, 1923–1958.

———, *The 'Sphere' of Sacrobosco and Its Commentators.* Chicago: University Press, 1949.

Turbayne, Colin M., "Grosseteste and an Ancient Optical Principle," *Isis* 50 (1959), pp. 467–472.

Tursman, Richard, ed., *Studies in Philosophy and the History of Science:* Essays in Honor of Max Fisch. Lawrence, Kans.: Coronado Press, 1970.

Valsanzibio, Silvestro da, *Vita e dottrina di Gaetano di Thiene, filosofo dello studio di Padova 1387–1465*, Verona: Scuola Tipografica Madonna di Castelmonte, 1948; 2d ed., Padua: Studio Filosofico dei Fratrum Minorum Cappuccini, 1949.

Van Leeuwen, Henry G., *The Problem of Certainty in English Thought 1630–1690.* With a Preface by Richard H. Popkin. The Hague: Martinus Nijhoff, 1963.

Villoslada, Ricardo G., *La Universidad de Paris durante los estudios de Francisco de Vitoria, O.P., 1507–1522*, Analecta Gregoriana, Vol. XIV. Rome: Gregorian University Press, 1938.

° Wallace, William A., "Albertus Magnus," *Dictionary of Scientific Biography*, Vol. 1, pp. 99–103.

° ———, "Aquinas, Thomas," *Dictionary of Scientific Biography*, Vol. 1, pp. 196–200.

———, "Buonamici, Francesco," *Dictionary of Scientific Biography*, Vol. 2, pp. 590–591.

———, "The 'Calculatores' in Early Sixteenth-Century Physics," *The British Journal for the History of Science* 4 (1969), pp. 221–232.

———, "The Concept of Motion in the Sixteenth Century," *Proceedings of the American Catholic Philosophical Association* 41 (1967), pp. 184–195.

266 Bibliography

Wallace, William A., "Coronel, Luis Nuñez," *Dictionary of Scientific Biography*, Vol. 3, pp. 420–421.

——, "The Enigma of Domingo de Soto: *Uniformiter difformis* and Falling Bodies in Late Medieval Physics," *Isis* 59 (1968), pp. 384–401.

——, "Galileo and the Thomists," to appear in *St. Thomas Aquinas Commemorative Studies (1274–1974)*, Pontifical Institute of Mediaeval Studies, Toronto, Canada.

——, "Mechanics from Bradwardine to Galileo," *Journal of the History of Ideas* 32 (1971), pp. 15–28.

° ——, *The Scientific Methodology of Theodoric of Freiberg.* Studia Friburgensia, N.S. no. 26, Fribourg: University Press, 1959.

° Wartofsky, Marx W., *Conceptual Foundations of Scientific Thought:* An Introduction to the Philosophy of Science. New York: Macmillan, 1968.

Webster, Charles, "Harvey's *De generatione:* Its Origin and Relevance to the Theory of Circulation," *British Journal for the History of Science* 3 (1967), pp. 262–274.

Weisheipl, James A., "Albert the Great," *New Catholic Encyclopedia*, Vol. 1, pp. 254–258.

——, "Albertus Magnus and the Oxford Platonists," *Proceedings of the American Catholic Philosophical Association* 32 (1958), pp. 124–139.

° ——, *The Development of Physical Theory in the Middle Ages,* New York: Sheed & Ward, 1959; reprinted Ann Arbor: University of Michigan Press, 1971.

——, "Motion in a Void: Averroës and Aquinas," to appear in *St. Thomas Aquinas Commemorative Studies (1274–1974)*, Pontifical Institute of Mediaeval Studies, Toronto, Canada.

——, "Ockham and Some Mertonians," *Mediaeval Studies* 30 (1968), pp. 163–213.

——, "The Place of John Dumbleton in the Merton School," *Isis* 50 (1959), pp. 439–454.

——, "Repertorium Mertonense," *Mediaeval Studies* 31 (1969), pp. 185–208.

Westfall, Richard S., "The Development of Newton's Theory of Color," *Isis* 53 (1962), pp. 339–358.

——, *Force in Newton's Physics.* The Science of Dynamics in the Seventeenth Century. New York: American Elsevier, 1971.

—— "The Foundations of Newton's Philosophy of Nature," *The British Journal for the History of Science* 1 (1962), pp. 171–182.

——, "Newton, Isaac," *New Catholic Encyclopedia*, Vol. 10, pp. 424–428.

——, "The Science of Optics in the Seventeenth Century," [Essay review of A. I. Sabra's *Theories of Light*], *History of Science* 6 (1967), pp. 150–156.

———, "Uneasily Fitful Reflections on Fits of Easy Transmission," *The* ANNUS MIRABILIS *of Sir Isaac Newton 1666–1966,* ed. Robert Palter, Cambridge, Mass.: The M.I.T. Press, 1970, pp. 88–104.

Whittaker, Sir Edmund, *Space and Spirit.* Theories of the Universe and the Arguments for the Existence of God. Edinburgh: Thomas Nelson, 1946.

Whitteridge, Gweneth, *William Harvey and the Circulation of the Blood.* New York: American Elsevier, 1971.

Wiener, P. P., "The Tradition Behind Galileo's Methodology," *Osiris* 1 (1936), p. 733 ff.

Wilkie, J. S., "Harvey's Immediate Debt to Aristotle and Galen," *History of Science* 4 (1965), pp. 103–124.

Wilson, Curtis A., "From Kepler's Laws, So Called, to Universal Gravitation: Empirical Factors," *Archive for History of Exact Sciences* 6 (1970), pp. 89–170.

———, *William Heytesbury: Medieval Logic and the Rise of Mathematical Physics,* Madison: University of Wisconsin Press, 1960.

Zilsel, Edgar, "The Origins of William Gilbert's Methodology," *Journal of the History of Ideas* 2 (1941), pp. 1–32.

Index

absolute things (*res absolutae*), 54, 57, 126

abstraction, 68, 69, 76, 78, 80, 85

acceleration, cause of, 107, 111, 159, 179, 180; *see also* falling body; mean-speed theorem; motion

accident, 17, 101, 102; absolute, 126, 132; fixed, 125; relative, 126, 132; successive or flowing, 125; *see also* substance

Achillini, Alessandro (1463–1512), 144, 151, 239, 251; *Quatuor libri de orbibus*, 151

action and reaction, 61

Adam, C., 244, 254

agnosticism, 159, 160, 176, 182, 209

air, 149, 150, 195

Albert the Great (ca. 1200–1280), 10, 56, 65, 66–71, 74, 76, 77, 80, 87, 94, 98, 99, 101, 102, 115, 117, 133, 213, 225, 229, 231, 233, 251; on the *De caelo*, 226; on *De mineralibus*, 70; on the *Ethics*, 226; on the *Metaphysics*, 68, 225; on the *Meteorology*, 70, 225, 226; on the *Physics*, 68, 70, 225, 226; on the *Posterior Analytics*, 66–68, 79, 225

Albert of Saxony (1316?–1390), 104, 109–111, 113, 115, 130–133, 251; on the *De caelo*, 232; on the *Physics*, 109–110, 232

Albertus Magnus; *see* Albert the Great

Alcalá, University of, 136

alchemy, 2

Alexander, H. G., 255

Alexander of Aphrodisias (ca. 200), 67, 141

Alfarabi (870?–950), 67

Alfragani (9th c.), 84

Alhazen (965?–1039?), 51, 63; *Perspectiva*, 51, 95

Alkindi (d. 873?), 34

alteration, 105, 110, 133

amici Platonis, 68; *see also* Platonist(s)

Amico, Gianbattista (d. 1538), 152

analysis and synthesis, 194, 198; *see also* resolution and composition

Anaxagoras (499?–428 B.C.),
    46
Andrade, E. N. da C., 247,
    257
anima mundi, 94, 154; *see
    also* world soul
annihilation, 125
an sit, 191, 214, 246
a posteriori, 51, 75, 84, 142,
    151, 153, 189, 207; *see also*
    demonstration
Apostle, H. G., 215, 252
appearances, stellar, 83, 86,
    87, 106; *see also* save the
    appearances
a priori, 51, 75, 84, 98, 151;
    see also demonstration
Aquinas, Thomas (1225?–
    1274), 46, 71–88, 93, 94,
    102, 104, 108, 115, 117, 134,
    136, 143, 148, 151, 178,
    204, 213, 226, 228, 231,
    234, 237, 239, 243, 251;
    Contra Gentiles, 227; *De
    motu cordis*, 72; *De occultis
    operibus naturae*, 72, 229;
    on the *De anima*, 72; on the
    De caelo, 72, 81–88, 227;
    on the *De generatione*, 72;
    on the *De memoria et rem-
    iniscentia*, 72; on the *De
    sensu et sensato*, 72; on the
    Metaphysics, 227; on the
    Meteorology, 72; on the
    Physics, 72, 216, 226; on
    the *Posterior Analytics*, 72–
    80, 121, 221, 226, 227,
    237; on the *Sentences*, 228;
    Summa theologiae, 72, 87,
    88, 227, 228
Archimedes (287?–212 B.C.),
    19, 139, 150, 177, 252;
    Measurement of a Circle,
    220
archtypal ideas, 169; *see also*
    Kepler, Johannes; Plato
Aristoteles Latinus, 213, 214
Aristotelianism, 36, 51, 65,
    139, 169, 194; heterodox,
    117; orthodox, 117; scho-
    lastic, 127, 139
Aristotelians, 133, 136, 144,
    151, 178, 179, 184–186;
    Averroist, 98, 121, 127,
    139, 149
Aristotle (384–322 B.C.), vi, 10,
    11–18, 20, 21, 28, 30–32,
    34, 40–44, 49, 53, 55, 56,
    58, 59, 63, 65, 66, 70, 73,
    75–79, 81, 84, 86, 95–97,
    100, 102, 103, 106, 107, 109,
    113, 117, 119, 121, 123,
    124, 127, 131, 133, 140–142,
    154, 160, 170, 177–179,
    184–186, 193, 213, 216,
    225, 243, 245, 247, 252;
    Categories, 15; *De caelo*,
    17, 21, 168, 228; *De so-
    phisticis elenchis*, 221; *Meta-
    physics*, 168, 215, 217; *Me-
    teorology*, 17, 18, 44, 46,
    142, 217; *On the Genera-
    tion of Animals*, 216; *Phys-
    ics*, 10, 11, 14, 17, 28, 29,
    58, 60, 68, 69, 75, 103,
    105, 119, 131, 132, 137,
    144, 152, 168, 185, 214–
    216, 247; *Posterior Analyt-
    ics*, 10, 11, 13, 17, 20, 28,
    29, 44, 46, 48, 53, 63, 64,
    66, 67, 72, 95–97, 118–
    120, 123, 154, 160, 168,
    176, 178, 185, 192, 194,
    213, 216, 246; *Prior Analyt-
    ics*, 11
arithmetic, 37
arrow, shot upward, 109, 111
articulus Parisienses, 105, 110,
    125; *see also* condemna-
    tions; Tempier, Etienne
astrolabe, 50, 90
astrology, 2, 112
astronomical tables, 151
astronomy, 47, 81, 151–153,
    162, 192
atomism, 19, 183, 184, 244

attraction, 172, 208; gravitational, 171, 208; law of, 91, 154, 209; magnetic, 91, 163, 169, 171, 173; mutual, 172
augmentation, 133
Augustine, Saint (354–430), 117, 218
Augustinianism, 51, 65
Augustinian(s), 121, 132
authority, argument from, 178
Averroës (1126–1198), 49, 66–68, 72, 117, 120, 121, 124, 134, 140, 225, 234, 237, 243, 253; on the *Physics*, 215, 234; on the *Posterior Analytics*, 121
Averroism, 94, 118, 177; Latin, 117, 225
Averroists, Renaissance, 152, 237
Avicenna (980–1037), 40, 42, 66, 68, 117, 225, 229
Aymeric de Plaisance (fl. 1304), 94

Bacon, Francis (1561–1626), 161, 162, 190
Bacon, Roger (1219?–1292?), 10, 28, 29, 47–52, 58, 63, 68, 71, 88–90, 93, 99, 102, 219, 221, 222, 225, 230, 253; *Communia naturalium*, 227; *De multiplicatione specierum*, 227; *Opus maius*, 47, 49, 90, 221, 227, 229, 230; *Opus tertium*, 225, 227, 228
Barrow, Isaac (1630–1677), 195, 249
Baumgardt, C., 240, 257
Baur, L., 217, 220, 254
Bellarmine, Robert (1542–1621), 88
Beltrán de Heredia, V., 236, 257
Benedetti, Giovanni Battista (1530–1590), 150, 239, 253

Berkeley, George (1685–1753), 162
Bernard, Claude (1813–1878), 160, 162
Bernard of Verdun (fl. 1280), 87, 228, 253
Blackwell, R. J., 226, 252
blood, 185; circulation of, 186–188, 191, 206; quantity of, 187–189
Boehner, P., 223, 256
Boethius (480–525), 213
Boethius of Dacia (fl. 1270), 117
Bohm, D., 212, 257
Borgnet, A., 67, 225, 251
Born, M., 212, 257
Borro, Girolamo (1512–1592), 149, 150, 253; *De motu gravium et levium*, 149–150, 178, 238
Boyer, C. B., 224, 230, 257
Bradwardine, Thomas (1290?–1349), 53, 58–59, 61–63, 104, 121, 130, 223, 253; *Tractatus de continuo*, 58; *Tractatus de proportionibus velocitatum*, 58
Brahe, Tycho (1546–1601), 168, 170–172, 241
Bridges, J. H., 221, 253
Brody, B. A., 211, 257
Bromberger, S., 211, 257
Bruno, Giordano (1548–1600), 154, 168
Bryson of Heraclea (4th c. B.C.), 38, 220
Buchdahl, G., 212, 249, 257
Bunge, M., 212, 258
Buonamici, Francesco (1540?–1603), 177, 179, 242, 253; *De motu libri decem*, 177, 178
Buridan, Jean (1295?–1358?), 103, 104–111, 113, 115, 121, 130, 132, 134, 230, 231, 253; on the *De caelo*, 232; on the *Ethics*, 231; on

Buridan, Jean (*continued*)
the *Metaphysics*, 231, 232;
on the *Physics*, 230–232
Burke, R. B., 221, 253
Burley, Walter (1275?–1345?),
55–58, 126, 132, 134, 223,
253; on the *Physics*, 56; on
the *Posterior Analytics*, 56;
Tractatus primus, 223; *Tractatus secundus*, 223
Burtt, E. A., 170, 212, 241,
243, 249, 258
Butts, R. E., 248, 249, 258

Cairns, H., 256
Cajori, F., 240, 249, 255
Calculator, 60; *see also* Swineshead, Richard
calculators, 118, 124, 130, 237
calculatory techniques, 121,
126, 148
Calippus (4th c. B.C.), 86
Callus, D., 218, 258
Cambridge: University, 194,
195; Platonists, 249
Capuano of Manfredonia, Francesco (fl. 1495), 152, 239,
253; on the *Theoricae novae
planetarum*, 239, 253
Carugo, A., 242
Caspar, M., 240–242, 258
Cassirer, E., 244
category, 17, 68, 136
causal: analysis, 15, 22, 189,
191; definition, 43, 189; explanation, vi, 5, 8, 13, 23,
39, 105, 114, 160–162,
176, 179, 181, 182, 184,
190, 196, 207, 210, 242,
244; mechanism, 184
causality, 5, 9, 55, 58, 62,
66, 160, 161, 164, 177,
210; and scientific explanation, vii, 7, 12, 62, 64, 210
passim; deterministic concept
of, 24
cause, 11, 132, 136, 137, 143,
144, 147, 148, 173, 190,
204, 215, 246; archtypal,
170, 174; common, 165, 173;
fictitious, 160; first, 32, 244;
immediate, 246; incidental,
40, 85, 86; mathematical, 41,
183; mechanical, 161, 184,
209, 210; occult or hidden,
160, 165, 183, 210; physical or natural, 39, 41, 64,
84, 85, 168–170, 230;
proper, 34, 37, 40, 180, 193,
197, 198; proximate, 30, 193,
222, 250; remote, 222; superfluous, 205; ultimate, 193;
see also efficient cause; final
cause; formal cause; material cause; true cause
cause and effect, 51, 123–125,
127, 140, 144, 148, 151,
182, 243, 244; convertibility of, 33; necessary relationship between, 122, 144,
145, 152
causes, 8, 70, 92, 97, 104, 114,
115, 141, 154, 160, 174,
181, 182, 186, 205, 231,
243; Aristotle's four, 13, 43,
72, 78, 101, 144, 170, 177,
204; knowability of, 161;
search for, 6, 23, 29, 47,
64, 95, 160, 168, 178, 184,
192, 193, 206
Celaya, Juan de (1490?–1558),
135–136, 253; on the
Physics, 135, 236
censure, ecclesiastical, 115; *see
also* condemnations
certitude, 29, 87, 120, 121,
141, 143, 188, 202, 250
chance, 16, 75
Charlton, W., 215, 252
circular reasoning (*circulatio*),
123, 124, 140
Clagett, M., 107, 130, 224,
232, 233, 235, 256, 258
classical science, vii, 21, 156–
210; founders of, 23, 148,
159–210; methodologists of,

23, 161, 162; philosophers of, 23, 161, 162; precursors of, 6, 95, 149
Clavelin, M., 245, 258
Cohen, I. B., 247, 249, 255, 258
color, 100, 193–195, 205, 247; formation of, 100, 196–198; nature of, 200, 208; reality of, 102
Columbus, Christopher (1451–1506), 228
combustion, 50, 206–207
comets, 18
Commentator, 124, 134; *see also* Averroës
composition, 119, 121, 127, 130, 140, 145; *see also* resolution and composition
Comte, Auguste (1798–1857), 162
Conciliator, 118; *see also* Pietro d'Abano
condemnations: of 1270, 103, 117; of 1277, 103, 105, 117
conjecture, 141–143; necessary, 143; rhetorical, 143
connotation, 136
consistency, 3
constructions, imaginary mathematical, 103
contemporary science, vii, 21
continuity, methodological, vii, 23
continuum, 16, 19
contradiction, 3
Conway, P. H., 252
Copernicus, Nicholas (1473–1543), 83, 151–154, 168–170, 228, 239, 241, 254; *On the Revolutions of the Celestial Spheres*, 83, 153, 167
Cornford, F. M., 215, 252
Coronel, Luis Nuñez (d. 1531), 133–135, 235, 254; *Physice perscrutationes*, 133–135, 236
cosmology, 163

Cotes, Roger (1682–1716), 249
Crew, H., 239, 254
Crombie, A. C., 28, 42, 94, 212, 217–221, 226, 228–230, 232, 242, 258
Crosby, H. L., Jr., 253
cumulative growth of knowledge, v, 193
Cyprus, 83

Dagut, M., 217
Dales, R. C., 218–220, 254, 258
Davis, J. W., 248, 258
declination, magnetic, 165
deduction, 16, 183; see also *a priori*; demonstration
deferent, 134; *see also* epicycle
definition, 16, 20, 43, 44, 60, 73, 96, 119, 189; as related to demonstration, 43, 67, 78, 97
Demiurge, 170
Democritus (5th c. B.C.), 19
demonstration, 16, 30, 43, 44, 73–76, 80, 104, 119, 140, 143–146, 152, 165, 185, 187, 189, 190, 193, 196, 202–204, 233, 234; absolute (*simpliciter*), 122, 142, 143, 153; *a posteriori*, 142, 151, 207; astronomical, 34, 35; circularity in, 123, 140; conjectural (*coniecturalis*), 142, 143, 148; mathematical, 121, 143; perfect, 31; physical, 121, 141, 143; relation to definition, 43, 67, 78, 97; see also *ex suppositione*; proof; *propter quid*; *quia*; sign
Denomy, A. J., 233, 256
De Salvio, A., 239, 254
Descartes, René (1596–1650), 95, 114, 161, 162, 182, 184, 190, 193, 195–197, 201, 244, 247, 254
determinism, 23, 161

dialectica, 31
dialectics, 17, 81, 86, 98, 148
Dijksterhuis, E. J., 223, 230, 259
Dikshoorn, C., 223, 259
dip, magnetic, 165, 166; cause of, 166–167
discovery, 98, 119, 127, 142, 155, 163, 181, 187; method of, 120, 145, 146; *see also* causes, search for
distance, 111, 127; *see also* velocity
distinction, 136; modal, 136; of reason, 136; real, 136
doctrina: compositiva, 119; *resolutiva*, 119
Dod, B. G., 214, 262
Doland, E., 212, 259
Dominican(s), 66, 68, 88, 153
Drabkin, I. E., 239, 242, 243, 254, 259
Drake, S., 182, 239, 242, 244, 254, 259
Duhem, P., 103, 104, 107, 108, 133, 137, 151, 153, 212, 217, 228, 230, 236, 239, 259
Dullaert of Ghent, Jean (1470?–1513), 132–133, 138, 254; on the *Physics*, 235
Dumbleton, John (fl. 1345), 53, 58, 59, 130, 223
Duns Scotus, John (1266?–1308), 106, 117, 134
Dyck, W. von, 241, 255
dynamics, 53, 63, 103, 127, 128, 130, 150, 245, 247

earth, 163, 165, 171; center of, 171, 172; circumference of, 84; pierced through center, 113; translational motion of, 113; rotation of, 108, 109, 111, 113, 154, 171, 242; shape of, 86, 166; sphericity of, 81, 83–85, 93, 219, 228

Eastwood, B. S., 218–221, 259
eccentric, 86, 87, 108, 134, 151, 152; and epicycle, reality of, 152, 153, 228
eclecticism, 94, 132, 137, 177
eclipse, 31, 33, 38, 67, 76, 122; lunar, 32, 73–75, 82, 83, 123, 192, 219
Edwards, W. F., 238
effect, 55, 125, 137, 140, 145, 183, 200, 207; *see also* cause and effect
efficient cause, 13, 14, 44–46, 48, 62, 63, 74, 75, 79, 98, 101, 168, 170, 171, 174, 175, 178, 186, 189–191, 193, 214, 215, 244, 246; *see also* cause
Egypt, 83; Egyptians, 18
Einstein, Albert (1879–1955), 3, 211
elements, theory of the, 100
Elie, H., 131, 235, 255, 259
Empedocles (492?–432? B.C.), 46
empirical: approach, 79, 115, 207; findings, 68; orientation, 22, 149; reasoning, 70, 154
empiricism, 6, 102, 161, 170, 192
end, 73, 75, 79, 189, 190; *see also* final cause; teleology
energy, 16, 175; magnetic, 167
ens: rationis, 5, 136; *reale*, 57
entity: abstracted, 69; fictional, 5; permanent, 56, 57, 132; relative, 55; successive, 56, 57, 105, 131, 132
epicycle, 87, 108, 134, 151, 152, 170; *see also* deferent; eccentric
error Platonis, 68
essence, 15; question of (*quid sit*), 12; see also *quid*; quiddity

ether, 210
ethics, 194
Euclid (fl. 295 B.C.), 34, 254;
    Catoptrica (attributed), 38
Eudoxus (400?–347? B.C.), 86
Evans, M. G., 213, 259
evidence, sense, 143
example, imaginary, 64, 127–
    129, 138, 224
exemplarism, 28
existence, question of (*an sit*),
    12; real, 54
experience, 43, 47, 49, 68,
    133, 149, 177, 181, 183,
    185, 207, 238
experientia, 43, 107, 149, 244;
    see also experiment; *experi-
    mentum*
experiment, 29, 71, 92, 100,
    133, 150, 163, 166, 181,
    188, 196, 197, 201, 202,
    206–208, 238, 239; cru-
    cial, 197; demonstrative,
    202, 203; imaginary, 52,
    182, 244; repeated, 71, 150,
    183, 199; see also *experi-
    mentum crucis*
experimental: apparatus, 151;
    method, 88–103, 118, 149,
    181; science or philosophy,
    29, 47, 49, 94, 98, 149,
    204, 206, 209; test, 150, 195
experimentalism, 47, 163
experimentation, 22, 41–43,
    50–52, 63, 64, 92, 95, 115,
    129, 130, 149, 181, 183,
    192, 196, 206; controlled,
    98
experimentum, 43, 88, 107,
    149, 150
experimentum crucis, 196–205,
    248
explanation, 1, 2, 12, 62, 81,
    159, 170; complete, 190,
    191; economy of, 108; hy-
    pothetical, 151, 154, 196,
    201; logical, 4, 5; mathe-
    matical, 18, 22, 27, 62, 88,

244; mechanistic, 112, 200,
    210; ontological value of,
    24; peripatetic, 210; physi-
    cal, 21, 27, 88; provisional,
    152; real, 21; scientific, vi,
    1, 2, 4, 5, 12, 76, 77 pas-
    sim; teleological, 16, 220;
    ultimate, 193, 244, 250;
    see also causal explanation;
    causality and scientific ex-
    planation
explanatory factors, 14, 216;
    see also causes
exponents, 112; irrational, 112,
    233; rational, 112, 233
ex suppositione, demonstration,
    75, 76, 79, 80, 102, 104, 143
eye, structure of, 52

Fabricius of Aquapendente, Hi-
    eronymus (1533?–1619), 186,
    188, 245
fact, 11, 12, 133, 182, 203;
    knowledge of, 12, 29, 86,
    191; question of (*quia*), 12
faith, 103, 113, 115, 232
fallacia consequentis (fallacy
    of the consequent), 41
falling body, 61, 64, 85, 106,
    111, 113, 137–139, 150,
    179–180, 206, 224, 236
falsification, 17, 29, 31, 39,
    41, 93, 101, 197
Favaro, A., 176, 178, 242, 254
fever, causes of, 126
Feyerabend, P. K., v, vi, 248,
    259
Ficino, Marsilio (1433–1499),
    154
final cause, 13, 14, 16, 46, 62,
    72, 73, 75, 78, 79, 101,
    102, 104, 144, 170, 189–
    191, 214, 215; extrinsic and
    intrinsic, 79, 227; *see also*
    cause
fire, 49, 50, 122, 133, 205
Fisch, M., 245, 265
Florence, 149, 176

fluxus, 105, 110; *fluxus formae*, 57, 68, 144

force, 58, 61, 103, 107, 115, 150, 170, 172–174, 176, 208, 247, 249; animating, 170, 171; attractive, 173; central, 173, 175; directive, 165; gravitational, 172; impressed, 106, 107, 195; magnetic, 90, 165, 167, 168, 173, 240; motive, 128, 170, 175, 178, 179; of impact, 107

form, 18, 19, 45, 69, 80, 101, 161, 163–165; mathematical, 80, 85; prime, 163, 164; substantial, 164, 247; *see also* formal cause; forms

forma fluens, 57, 68, 105, 144; *forma remissa*, 144

formal cause, 13, 14, 43–45, 78, 101, 164, 168, 170, 174, 175, 189, 214, 215; *see also* cause

formalistic analysis, 7, 21

forms, 19, 20; intension and remission of, 56, 57, 60, 126, 133, 135, 206, 207; latitude of, 59, 60, 128, 135

Foscarini, Paolo Antonio (1580–1616), 88

Fowler, T., 246

Fracastoro, Girolamo (1483–1533), 152, 154

Franciscan(s), 28, 55, 106, 117, 222

Franciscus de Marchia (fl. 1320), 105–107, 231; *Reportatio*, 231

frequent occurrence, events of, 33, 74, 75, 122, 123

Frisch, C., 241, 255

Gaetano da Thiene (1387–1465), 118, 127–130, 138, 139, 235, 254; on the *Physics*, 235; on the *Regule*, 127–130, 235

Galen (130?–200?), 42, 119, 121, 225, 229, 234, 245; *Microtegni*, 234; *Tegni*, 119

Galenists, 185

Galileo Galilei (1564–1642), 9, 10, 19, 22, 29, 88, 107, 116, 118, 130, 137, 144, 149, 150, 159, 160, 162, 168, 176–185, 194, 195, 232, 233, 239, 242–244, 254; precursors of, 10, 104; *The Assayer*, 243, 244; *Juvenilia*, 176, 194; *On Motion*, 177–180, 182, 194, 238, 244; *Quaestiones logicae*, 176, 242; *Two Chief World Systems* (*Dialogi*), 180, 182, 243, 244; *Two New Sciences* (*Discorsi*), 159, 182, 183

Gaye, R. K., 215, 252

genus, 17

geometrism, 47

geometry, 37, 38, 40, 48, 78, 97, 154, 175, 178, 181; analytical, 114; configurational, 113, 114; non-Euclidean, 4; projective, 82

George of Brussels (fl. 1495), 133, 134

Gerard of Cremona (1114?–1187), 213–215

Geymonat, L., 184, 242–244, 259

Gilbert, N. W., 234, 259

Gilbert, William (1540–1603), 94, 154, 160, 162–169, 171, 176, 183, 229, 240, 254; *De magnete*, 163–167, 172, 229, 240; *De mundo*, 163

Gilbert de la Porrée (1075?–1154), 56; *Liber de sex principiis*, 56

Giles of Lessines (1235?–1304?), 226

Giles of Rome (1247?–1316), 121, 254; on the *Physics*, 215; on the *Posterior Analytics*, 121

Gillispie, C. C., 259

Gilson, E., 218, 229, 259
Girill, T. R., 244, 259
Glanville, J. J., 234, 237, 260
God, 53–54, 107, 169–170; absolute power of, 53, 103, 105; conserving action of, 54
Grant, E., 224, 232, 256, 260
gravitation, 8, 85, 206; law of universal, 207–209; mutual, 207, 208, 249; tendency toward, 85, 171
gravity, 106, 107, 115, 154, 160, 171, 173, 182, 194, 195, 207, 208, 249; cause of, 160, 193, 205–210; center of, 84, 113, 150; explanation of, 199, 209–210; reality of, 208, 210; specific, 150
Greek science, 8, 11, 22
Gregory of Rimini (d. 1358), 125, 132–134, 230–231
Grimaldi, Francesco Maria (1618–1663), 199, 201
Grosseteste, Robert (1168?–1253), 10, 27, 28–53, 56, 58, 59, 62, 65–68, 71, 72, 74, 77–79, 82, 87, 98, 101, 114, 115, 119, 154, 213, 218, 229, 231, 254; *De artibus liberalibus*, 221; *De cometis*, 41; *De generatione sonorum*, 221; *De iride*, 37, 39, 96, 221; *De libero arbitrio*, 218; *De lineis*, 30, 218; *De luce*, 35; *De sphaera*, 33, 35, 219; on the *Physics*, 28, 29, 218; on the *Posterior Analytics*, 29–46, 48, 67, 86, 218
Guthrie, W. K. C., 228, 252

Hall, A. R., 256
Hall, M. B., 250, 256, 260
halo, 95
Hamilton, E., 256
hand, invisible, 249
Hardie, R. P., 215, 252

harmony, 37, 169; dynamic, 170; mathematical, 168–170
Harré, R., 212, 260
Harvey, William (1578–1657), 10, 162, 184–193, 194, 206, 245–247, 255; *Generation of Animals*, 185, 186, 245; *Motion of Animals*, 245; *Motion of the Heart and Blood*, 185, 190, 192, 245, 246; *Second Disquisition to Riolan*, 190, 193
heart: definition of, 189; motion of, 185, 186, 188, 190, 246; pumping action of, 193, 246
heat, 122, 126, 134, 170, 190; motion as the cause of, 244
Heath, T. L., 220, 252
heavens: atmospheric region of, 101; conjunctions and oppositions in, 112; influence on magnet, 93; motion of, 106, 112; ultimate sphere of, 105, 125; *see also* appearances, stellar; moon; planet(s); sun; universe
Heiberg, J. L., 254
Heisenberg, W., 3, 211, 260
Hellman, C. D., 240, 258
Hellmann, G., 229, 256
Hempel, C. G., 211, 260
Henry of Ghent (1217?–1293), 117
Herberstein, Baron von (fl. 1596), 241
Herivel, J., 194, 247, 260
hermetic tradition, 93
Hero of Alexandria (fl. 60), 150
Herschel, John F. W. (1792–1871), 162
Hesse, M. B., 212, 260
Heytesbury, William (1313?–1372?), 53, 58–60, 127–130, 133, 255; *Regule*, 127–129
Hipparchus (fl. 135 B.C.), 86
Hippocrates (460?–377? B.C.), 71

history of science, vi, vii, 65;
approach of, 7, 9
Hobbes, Thomas (1588–1679),
162
Holkot, Robert (1290?–1349),
230
Hollenhorst, G. D., 234, 255
Holtsmark, T., 248, 260
Hooke, Robert (1635–1703),
199–201, 210, 247
Hope, R., 215, 216, 252
Hoskin, M. A., 224, 257, 262
Hugo of Siena (d. 1439), 126,
127; on the *Tegni*, 234
humanism, Renaissance, 154
human soul, 154, 240; immortality of, 139, 237
Hume, David (1711–1776),
5, 7, 162
Huygens, Christian (1629–
1695), 199, 200, 210; *Discourse on the Cause of Gravity*, 210
hypothesis, 86, 142, 143, 151,
153, 160, 169, 197, 200,
201, 206, 208, 210, 225;
mechanical, 199–201, 209,
210
hypothetical: procedure or
method, v, 76, 196, 209–
210; proposition, disguised
hypothetical, 231; reasoning, 107, 108, 134, 141,
143, 148, 200; syllogism,
41, 141, 142

idea, 19, 169
illumination, divine, 28
impetus, 110, 113, 115, 134,
135, 154; curvilinear, 106,
107; impressed by gravity,
106; permanent, 110, 112;
quantified, 106; rectilinear,
106; self-expending, 106,
110, 112; theory of, 55,
106–107
incipit et desinit, 56
inclined plane, 130

incorruptible, 31, 32, 33, 67
indeterminacy, 3
induction, 12, 16, 100, 145,
146, 185, 193, 194, 206,
208, 209, 249; demonstra-
tive, 147; magnetic, 92
inertia: planetary, 174, 175;
principle of, 55
instrument, 92, 151
intellect, 140, 141, 147, 192
intelligibility, 147
intensification, 60, 61, 135,
206; *see also* forms, intension and remission of
iron, magnetized, 91

Jacopo da Forlì (d. 1461), 126,
134; on the *Tegni*, 234
Jaki, S. L., 212, 230, 260
James the Venetian-Greek (12th
c.), 213–215
Jammer, M., 176, 239, 241,
260
Jeffery, G. B., 211, 261
Jesuits, 199, 200–204
Joannes Maior; *see* Mair, Jean
John Damascene (645?–750?),
218
John Major of Haddington;
see Mair, Jean
John of Jandun (1275?–1328),
118
John of St. Thomas (1589–
1644), 234, 255
John of Salisbury (1115?–
1180), 213
John the anonymous translator
(12th c.), 213, 214, 220
John Tzetzes (12th c.), 217
Jung, C. G., 241, 260
Jungius, Joachim (1587–1657),
144

Kant, Immanuel (1724–1804),
5, 162, 249
Kepler, Johannes (1571–1630),
19, 154, 162, 167–176,
183, 194, 210, 240, 241,

243, 255; second law of, 172; third law of, 175; *Astronomia nova*, 171, 173, 241; *Epitome of Copernican Astronomy*, 168, 169, 172, 240, 241; *Mysterium cosmographicum*, 173, 243; *Tertius interveniens*, 241
Kilwardby, Robert (1215?–1279), 68, 69, 77, 87, 225; *De ortu scientiarum*, 68
kinematics, 53, 60, 103, 127–130, 132, 137, 245
Koyré, A., 210, 212, 244, 249, 250, 260–261
Krebs, E., 229, 261
Kren, C., 228, 261
Kuhn, T. S., v, vi, 211, 247, 248, 261

Larcher, F. R., 221, 252
law, 91, 160, 172; causal, 104; inverse-square, 209; mathematical or quantitative, 42, 116, 139, 172; of falling bodies, 113, 137, 207; of nature, 76, 161, 227, 243, 244
Leake, C. D., 246, 255
Lee, H. D. P., 217, 252
Leff, G., 217, 218, 224, 225, 230, 233, 261
Leibniz, Gottfried Wilhelm (1646–1716), 144, 161, 162, 210, 255; *Tentamen de motuum caelestium causis*, 210
Lenzen, V. F., 212, 261
Leonardo da Vinci (1452–1519), 107, 111
less known to nature, 124
less known to us, 121, 143
lever, 151
Liber de causis, 65
Liège, 202
light, 160, 195, 205; corpuscular theory of, 204; extramission theory of, 34; mechanistic theories of, 197; meta-

physics of, 28, 35, 43, 46, 48, 52, 53, 62, 77, 80, 93, 95, 230; nature of, 194, 200, 208, 209; refrangibility of, 203; self-diffusion of, 28, 35; white, 198, 199
lightning, 43, 44, 46, 67, 76, 225; *see also* thunder
Lincoln, Bishop of; *see* Grosseteste, Robert
Lindberg, D. C., 52, 219, 221, 222, 256, 261
Line, Francis (1595–1675), 202, 248
Litt, T., 228, 261
loadstone, 90–93, 163–167; *see also* magnet
Locke, John (1632–1704), 162
logic, 4, 104, 121, 144, 181, 187, 194; formal, 11, 21; material, 11, 104, 234; nominalist, 121; terminist, 104
London, 186
Lorentz, H. A., 211, 261
Lucas, Anthony (d. 1693), 202–204, 247, 248
Lynch, L. E., 218, 261

Maestlin, Michael (1550–1631), 168, 241
magic, 89, 93, 149
magnet, 90–94, 154, 162–167, 172; poles of, 90–92, 165, 166; reversal of poles of, 92; spherical, 93, 164; see also *terrella*
magnetic: faculty, 171; invisible chains, 171; phenomena, causes of, 163
magnetism, 162, 169
Mahoney, E. P., 237, 261–262
Maier, A., 105, 212, 225, 231, 232, 262
Mair, Jean (1469–1550), 131–133, 135, 255; on the *Physics*, 235
Manuel, F. E., 247, 262

Maricourt, 88; *see also* Peter of Maricourt

Marliani, Giovanni (d. 1475), 130, 235; *Quaestio de proportione motuum in velocitate*, 130

Maschler, C., 212, 259

mass, 16, 172

material cause, 13, 14, 16, 43–45, 48, 62, 74, 75, 79, 98, 101, 170, 186, 189, 204, 214, 215; *see also* cause; matter

materia prima, 16

mathematical: insight, 184; physics, 36, 59, 81, 196, 214; principles, 194; reasoning, 13, 47, 64, 93, 115, 118, 148, 154, 180, 181, 207

mathematicism, 52, 63, 78, 87, 93, 95, 170

mathematics, 4, 17, 20, 29, 47–49, 52, 69, 74, 80, 81, 88, 109, 114, 123, 142, 145, 175, 177, 181, 195

matter, 16, 18, 19, 45, 80, 81, 134, 163, 165, 176, 207; atomic theory of, 183; structure of, 194

Maximus the Confessor (580?–662), 218

Mazzolini, Sylvester de Prierio (1460–1523), 153, 239; on the *Theoricae novae planetarum*, 153, 239

McAllister, J. B., 229, 252

McMullin, E., 239, 262

mean-speed theorem, 60, 113, 137, 139

measurement, 63, 84, 101, 129

mechanics, 58, 115, 150, 247; celestial, 172; principles of, 194

medieval science, vii, 8, 21, 22, 27–116, 161; *see also* science

Menut, A. D., 233, 256

Merlan, P., 220, 262

Merton College, 27, 55, 104, 121

Mertonians, 27, 53–64, 109, 115, 127, 128, 137, 222, 223; ambivalence of, 125

metaphysics, 29, 65, 69, 74, 212, 249; *see also* light, metaphysics of

meteorological sphere, 99

meteorology, 37, 97

method, 144, 178, 202, 246; demonstrative, 145, 203; resolutive, 145

methodology, 72, 74, 81, 96, 149, 155, 178, 180, 182, 190, 196, 243; Aristotelian, 11–18, 185–189, 213; demonstrative, 204; scientific, 1, 17, 118, 149, 161, 206, 210

Michael Scot; *see* Scot, Michael

middle term, 12, 38, 81, 97, 127, 140, 141, 204

Mill, John Stuart (1806–1873), 162

Minio-Paluello, L., 214, 215, 262

Minkowski, H., 211

mirror, burning, 89

Mittelstrass, J., 217, 262

modern science; *see* classical science; science

Molland, A. G., 224, 257, 260

Moody, E. A., 109, 111, 230, 231, 253, 262

moon: crescent, 82, 219; gibbous, 82, 219; phases of, 13, 36, 49, 51, 77, 82, 143, 219; shape of, 33, 34, 35, 122, 142; waxing and waning of, 82, 122, 142; *see also* eclipse, lunar

More, Henry (1614–1687), 249

more known to nature, 120, 121, 124

more known to nature and to us, 120, 143
more known to us, 120
motion, 55, 58, 68, 80, 104, 126, 131–133, 136, 137, 144, 170, 200, 243; causes of, 62, 103, 105, 130, 167, 170, 173, 178, 193, 210; celestial, 112, 125; circular, 174, 175, 187, 188; curvilinear, 125; difform (nonuniform), 61, 125, 128, 129, 137, 138, 172, 174; difformly difform, 138; entitative status of, 135, 136; laws of, 209; local, 55, 57, 58, 62, 105, 106, 110, 125, 127, 128, 130–132, 134–136, 177; natural, 92, 159, 164, 178–180; Ockhamist or nominalist view of, 54–55, 104, 110, 128, 131; realist view of, 104, 108, 110; reality of, 57, 59, 61, 103, 105, 106, 109–110, 125, 128, 131, 134, 144, 208; rectilinear, 125; relativity of, 86, 108; terrestrial, 112; uniform, 55, 61, 125, 128, 129, 137, 138; uniformly accelerated, 207; uniformly difform, 138, 139
motions, incommensurability of celestial, 112
motor causality, 55, 125, 136, 173; see also efficient cause
Motte, A., 240, 255
multiplication of species, 48
Murdoch, J. E., 223, 224, 233, 262
Mure, G. R. G., 213, 214, 252

Nagel, E., 1, 212, 262
Nardi, B., 239, 262
Nash, L. K., 212, 262
natural philosophy, 130, 134, 179, 194; see also natural science; physics

natural science, 80, 121, 122, 130, 142, 143, 145, 148; see also physics
nature, 15, 37, 103, 138, 154, 155, 188, 193, 207, 247; book of, 180, 185; common, 54; essential, 78; physical, 80; question of (quid sit), 12; uniformity of, 206, 220; see also law of nature
necessary connection, 144, 145, 147; see also cause and effect
necessity, 102, 122, 123, 143; absolute, 80, 122, 189; aptitudinal, 123
needle, magnetized, 92, 165
negotiatio, 140, 141, 147, 148
Neoplatonism, 28, 35, 36, 72, 94, 117, 154, 167, 168
Neopythagorean(s), 154, 168, 249
Neurath, O., 212, 262
Newton, Isaac (1642–1727), 159, 160, 162, 175, 182, 194–210, 240, 247–249, 255; Lectiones opticae, 195; Mathematical Principles of Natural Philosophy, 159, 194, 205–206, 208, 240, 247, 249; Opticks, 194, 195, 248; Quaestiones quaedam philosophicae, 195; Wastebook, 194
Nicholas of Autrecourt (1300?–1350?), 104, 231
Nicholas of Cusa (1401–1464), 154, 168, 170
Nicoletti, Paolo; see Paul of Venice
Nifo, Agostino (1473–1546?), 118, 139–144, 145, 147, 148, 151, 176, 237, 239, 256; on the De caelo, 153, 239; on the Physics, 140–143, 237; on the Posterior Analytics, 140, 143, 237; Recognitio, 141, 142, 237

nominalism, 52, 53, 59, 62, 117, 130, 230
nominalists, 110, 128, 131–137, 144, 225
North, J. D., 221, 258
number, 19, 69; nature of, 217; theories of, 4

observation, 70, 88, 149, 169, 174, 185, 186, 192, 196, 206, 208
occult work of nature, 92; see also cause, occult or hidden; quality, occult
Ockham, William of (1285?–1347), 52, 53–57, 59, 60, 62, 103, 105, 110, 125, 132, 222, 230, 256; Reportatio, 223; Summa totius logicae, 223; Tractatus de successivis, 223
Oldenburg, Henry (1615?–1677), 196
olive: generation of, 75, 76, 79, 123; universal concept of, 76
omne quod movetur ab alio movetur, 55, 173; see also motor causality
Oppenheim, P., 211, 260
optics, 35, 38, 63, 94–103, 196, 207, 248; geometrical, 35, 37, 38, 45, 51, 53, 96, 97; physical, 35, 39
orbits: elliptical or oval, 174, 175; planetary, 171–173
Oresme, Nicole (1320–1382), 104, 111–114, 115, 130, 133, 135, 138, 233, 256; De proportionibus proportionum, 112, 233; Le Livre du ciel et du monde, 233; Questiones de spera, 233; Tractatus de configurationibus, 114
Osiander, Andreas (1498–1552), 153
Owen, G. E. L., 216, 263

Oxford, 27, 63, 68, 102–103, 109, 117, 121, 127, 130, 222, 225; University of, 10, 22, 27–64, 102, 103, 217
Oxford Platonists, 69, 225

Padua, 98, 139, 140, 151, 152, 194; School of, 118–130, 139–155, 180; University of, 22, 116, 117–130, 139–155, 176, 184
Pagel, W., 245, 263
Palter, R., 248, 249, 263
Pappus (fl. 285), 150
Paracelsus (1493–1541), 93
Pardies, Ignace (d. 1673), 200–202
Paris, 27, 65, 81, 104, 114, 115, 128, 225; University of, 10, 22, 64, 65–116, 117, 121, 130–139, 217
Paris Terminists, 66, 103–114, 115, 124, 127, 128
part, component, 14, 204
Passmore, J. A., 247, 263
Pauli, W., 241, 260
Paul of Venice (1369?–1428), 118, 121–127, 128–130, 132, 134, 139–141, 256; on the Physics, 124–126, 234; on the Posterior Analytics, 121–124, 234; Summa philosophiae naturalis, 124, 234
Pavia, University of, 130
Peckham, John (1240?–1292), 29, 47, 51, 52, 63, 219, 221, 256; Perspectiva communis, 51, 222
pendulum, 130
peregrinus, 88; see also Peter of Maricourt
periculum, 150; see also experiment; experimentum
peripatetic(s), 154, 160, 164, 195, 199, 204
perpetual motion machine, 90

per posterius, 31; *see also* demonstration; *a posteriori*
Perrett, W., 211, 261
per se nota, 45
Peter d'Ailly (1350–1420), 134
Peter of Maricourt (fl. 1269), 66, 88–94, 95, 102, 115, 164, 222, 228, 229, 256; *De magnete*, 88, 164
Peter of Spain (1210?–1277), 104, 231; *Summulae*, 231
phenomena, electric vs. magnetic, 163; *see also* save the phenomena
phenomenalism, 77
Philolaus (5th c. B.C.), 217
Philoponus, John (6th c.), 140, 141, 142, 214
Philosophical Transactions, 196, 199, 247, 248
philosophy, 208; mechanical, 195, 250; realist, 115; *see also* metaphysics; natural philosophy
philosophy of science, v, vi, vii, 7, 9, 24, 162, 247; analytical approach to, 7–9
physics, 37, 40, 81, 96, 97, 133, 148, 151, 182
Pietro d'Abano (1257–1315), 118–120, 121, 151, 233; *Conciliator differentiarum*, 118, 119, 234
pin-hole images, 52, 122, 222
Pisa, University of, 149, 176, 177, 194
planet(s), 171–175, 205, 207; locations of, 33, 34, 49, 172; non-twinkling of, 12, 34, 77
Plato (427–347 B.C.), 17, 18–21, 68, 69, 72, 73, 140, 155, 168–170, 256; error of, 68, 69; *Philebus*, 217; *Republic*, 217; *Timaeus*, 20, 70, 169–170, 217, 241
Platonism, 10, 11, 154, 169, 180, 184, 244
Platonist(s), 80, 249; *see also*

Cambridge; Oxford Platonists; *amici Platonis*
Platt, A., 216, 252
Plochmann, G. K., 245, 263
Pohle, W., 217, 263
Poinsot, Jean; *see* John of St. Thomas
Pomerans, A. J., 211, 260
Pomponazzi, Pietro (1462–1525), 139–140, 237; *Quaestio de regressu*, 237
Popkin, R. H., 265
Popper, K. R., 17, 263
Poppi, A., 233, 237, 263
Porphyry (233–304), 56; *Isagoge*, 56
positivism, 1, 6, 176, 182, 184; logical, 4
Post, H. R., vi
Powers, D'Arcy, 245
precession of the equinoxes, 86, 168
predicate, 12, 38
prediction, 23, 112, 161
Premuda, L., 233, 263
presence: actual, 204, 205, 230; virtual, 204, 205
Prime Mover, 170
principle(s), 69, 87, 100, 123, 125, 136, 143, 145, 146, 152, 153, 166, 173, 193, 207; of economy, 106
prism, 95, 196–198, 230
Proclus (410–485), 65
projectile: 105, 106, 110, 171, 178, 179; motion, cause of, 178–179, 194, 195
proof, 81, 87, 127, 140, 144, 148; mathematical, 17, 80; method of, 120, 144; *see also* demonstration
proper subject, 98, 101; *see also* material cause
property, 17, 96, 97, 199–201, 210; original or connate, 199
proportionality, geometrical, 58, 59

propter quid, 12, 29–31, 33–
  36, 38–41, 43, 46, 63, 67,
  75, 77, 78, 80, 81, 96, 97,
  101, 102, 115, 119, 122,
  124, 125, 140–142, 191,
  219, 220, 234, 246; *see
  also* demonstration
Ptolemaists, 152
Ptolemy (fl. 140), 34, 86; *Almagest,* 95, 152; *see also*
  universe, Ptolemaic system
  of
pulley, 151
Purbach, Georg (1423–1461),
  152–153, 239, 256; *Theoricae novae planetarum,*
  152
Pythagoras (fl. 550 B.C.), 18,
  19, 168
Pythagorean(s), 18, 19, 20, 46,
  79, 217

quadrivium, 35
quality, 54, 69, 105, 136, 199,
  200, 206; occult, 154, 164,
  209, 210; original, 200; permanent, 106
quantification, 54
quantitative modality, 85, 204
quantity, 54, 80, 169, 187;
  of matter, 16; physical, 80,
  81, 84
quantum theory, 2, 24
questions: Aristotle's four scientific, 12, 214; "how" questions, 1, 160; "why" questions, 1, 14, 15, 160; see
  also *an sit;* explanation, scientific; *propter quid; quia;
  quid*
quia, 12, 29–31, 33–36, 38–
  41, 67, 77, 96, 119, 122,
  124, 125, 141, 219, 234,
  246; *see also* demonstration
quid, 96, 97, 101, 147, 148,
  220
quiddity, 15, 101
quintessence, 207

quod, 40, 148, 214, 220
quod quid, 78

rainbow, 13, 17, 33, 36, 37,
  39, 49, 76–79, 94–102,
  222; causes of, 95–102;
  primary, 99–100, 230;
  properties of, 51; secondary,
  99–100, 230
raindrop, 52; individual, 98;
  magnified, 99
Randall, J. H., Jr., 127, 130,
  233, 234, 237, 238, 263
random event, 16
ratio, 126, 127, 133, 137; irrational, 112; qualitative, 59;
  quantitative, 127, 128; *see
  also* ratios
rationalism, 52, 77, 161, 245
ratios: of motions, Aristotle's
  rules for, 107; of speeds in
  motion, 58, 130, 133
Ratner, H. A., 245, 246, 263
real entities, 5, 154; *see also*
  entity
realism, 1, 5, 7, 9, 24, 52, 59,
  62, 102, 103, 108, 121, 124,
  212, 231, 245; moderate,
  130
realists, 110, 128, 129, 131,
  132, 134–137, 144
reality, 109, 128, 153, 201,
  211, 232; known by faith,
  103, 113; physical, 87; ultimate, 19; *see also* color,
  reality of; eccentric, and
  epicycle, reality of; motion,
  reality of
reason, human, 115, 192, 232
reason, question of (*propter
  quid*), 12
reasoned fact, knowledge of,
  12, 29, 86, 191
red bile, purgation of, 42, 43,
  67, 90, 229
reflection, 39, 49, 52, 205, 207;
  internal, 99, 100

refraction, 42, 49, 52, 99, 100, 197–199
Regiomontanus (1436–1476), 153
regressus, 140, 142, 146, 147, 237, 238, 242
Reilly, C., 248, 263
relative things (*res relativae*), 57, 126
relativity, theories of, 2, 24
Renaissance, 22, 109, 144, 155, 160
Renaudet, A., 235, 263
resistance, 58, 61, 106, 107, 115; arising from gravity, 107; arising from medium, 128
resolution and composition, 29, 60, 98, 119–121, 133, 140, 180, 194, 198, 243; logical vs. real, 126
resolutive procedure, 127, 130, 145, 146
respiration, 205, 206
Rheticus, Joachim (1514–1576), 153
rhetoric, 194
Riedl, C. C., 219, 255
Riolan, Jean (fl. 1648), 190; *Encheiridium anatomicum*, 190
Rizzi, B., 228, 263
Rosen, E., 153, 263
Rosenfeld, L., 199, 248, 264
Ross, W. D., 216, 252
Roth, F. G., 234, 264
Royal Society, 196, 200
rules of reasoning, Newton's, 205–208, 249

Sabra, A. I., 247, 264
Sacrobosco, John of (13th c.), 33, 219; *Sphere* of, 219
Salamanca, University of, 136–139
Sambursky, S., 217, 264
Santillana, G. de, 228, 264
save the appearances, 21, 81, 86–88, 106–109, 115, 151–153
save the phenomena, 103, 108, 113, 133, 148, 151, 169
Scaliger, Julius Caesar (1484–1558), 173
scammony, 42, 43, 67, 90, 229; *see also* red bile, purgation of
Schiavone, M., 237, 264
Schlund, E., 228, 229, 264
Schmitt, C. B., 149, 232, 233, 236, 237, 242–244, 264
science, 6, 11, 29, 30, 104, 142, 143; biological, 162; mathematical, 120; "new," 22, 29, 130, 154, 160, 184; perfect, 145; physical, 161, 162; subalternated, 31, 36–38, 67, 69, 81; subalternating, 31, 36–38, 78, 97
scientia, 6, 31
scintillation, 49
Scot, Michael (1175?–1235), 214, 215
Scott, T. K., 231, 264
Scotus, Duns; *see* Duns Scotus, John
Scriven, M., 211, 264
sense perception, 12, 146, 153, 185, 192, 207
Settle, T. B., 239, 264
Shapere, D., 249, 264
Shapiro, H., 223, 264
Sharp, D. E., 226, 264
Shea, W. R. J., 242, 244, 264–265
Siger of Brabant (1240?–1281?), 117
Siger of Foucaucourt (fl. 1269), 90
sign, demonstration of, 121, 152–153, 234
Simon, Y. R., 234, 255
simplicity, 205
Simplicius (6th c.), 21, 141, 228; on the *De caelo*, 228
simplification, 64, 138

simultaneity, 3
Singer, C., 245, 265
skepticism, 190, 250
Snow, A. J., 249, 265
solar system, 207, 208
Sommerfeld, A., 211
sophismata calculatoria, 54, 60, 137
Soto, Domingo de (1494–1560), 135–139, 178, 236, 257; on the *Physics*, 136–139, 236
Soto, Francisco de, 135; *see also* Soto, Domingo de
soul, 167, 173, 193, 240, 247; *see also* human soul; world soul
sound, generation of, 221
space, 19, 172, 211; imaginary, 128
species, 17, 144
spectrum: colors of, 193; shape of, 196–198
sphere, 85, 86, 137
spheres: heavenly, 154, 170; homocentric, 151
spherical flask, 95, 99
Spiazzi, R. M., 251
spirit(s), 154, 190, 191, 193, 246
Sprague, R. K., 216, 265
Stannard, J., 245, 265
statue analysis, 15; *see also* causes, Aristotle's four
steelyard, 150
Strong, E. W., 249, 265
subalternation of the sciences, 77; *see also* science, subalternated, subalternating
subject, 12, 38, 96; *see also* proper subject
subjectivity of secondary qualities, 183
sublunary region, 74, 75, 170
substance, 54, 69, 101, 200; inherence of accidents in, 125
sun, 169, 170, 172, 173, 175, 205–208

supposition, natural, 231
Swineshead, Richard (fl. 1340–1355), 53, 58, 60–62, 133, 224, 257; *Liber calculationum*, 60, 61, 224; *Tractatus de motu locali difformi* (attributed), 61
Sylla, E., 223, 262
system of the world, 208; *see also* universe

Tannery, P., 244, 254
Tartaglia, Niccolò (1500–1557), 150, 239
Tartarus, 79
teleology, 170; *see also* explanation, teleological; final cause
Tempier, Etienne (d. 1279), 103, 225
terrella, 164, 166
Themistius (317–387), 67, 140, 141, 213
Theodoric of Freiberg (1250?–1310?), 63, 66, 88, 94–102, 115, 120, 226, 230, 257; *De accidentibus*, 101; *De coloribus*, 100, 230; *De elementis*, 100; *De iride*, 94, 230; *De miscibilibus in mixto*, 100
theology: Augustinian, 28, 94, 169; Christian, 10; Neoplatonist, 94
theory, 116, 148; three-color, 102; four-color, 100, 102
Thomist(s), 135, 153, 225, 242
Thorndike, L., 219, 224; 226, 265
thought experiment, 139, 149, 182; *see also* experiment, imaginary
thunder, 43–46, 67, 76, 78, 79, 225, 231
time, 111, 127, 172, 211
tollendo tollens, 41
Tomitanus, Bernardinus (fl. 1570), 144

tradition: Aristotelian, 126, 168, 176; Averroist, 154; Galenic, 126; Ockhamist, 231; scholastic, 154; technological, 150, 181

trajectory of cannon balls, 150, 180

Trinity, Holy, 169

true cause, 23, 100, 102, 143, 148, 160, 163, 166, 180, 181, 184, 188, 194, 197, 198, 201, 204, 205, 209, 210, 243, 244; see also *vera causa*

true form, 169

true pole of the heavens, 92, 93, 166

truth, v, 4, 20, 136, 185, 201–204, 225

Tübingen, 168

Turbayne, C. M., 221, 265

Turnbull, H. W., 247, 255

Tursman, R., 245, 265

twinkling, 12, 35, 49, 51, 77

two hands, man with, 74–76

Ubaldo del Monte, Guido (1545–1607), 150, 239

Ulrich of Strassburg (d. 1278?), 226

uncertainty, 3

uniformiter difformis, 139, 236; see also motion, uniformly difform

universal, 32, 33, 67, 73, 74, 76–77, 146, 147, 231; conditioned, 33

universale experimentale, 42, 101

universe: center of, 61, 170, 171, 173; clockwork, 111; composed of homocentric spheres, 86, 151; Copernican system of, 88; Eudoxian system of, 86; geocentric, 86, 170; heliocentric, 167, 169, 170; Ptolemaic system of, 2, 86, 151; structure of, 194

universities, 27, 150; see also Alcalá; Cambridge; Oxford; Padua; Paris; Pavia; Pisa; Salamanca

Urban the Averroist (fl. 1334), 120–121; on the *Physics*, 120

vacuum or void, 139, 178, 243

Valsanzibio, S. da, 235, 265

Van Leeuwen, H. G., 250, 265

variation, magnetic, 165, 166

velocity, 127–129, 132, 133, 137, 139, 172; and distance of fall, 111, 137, 138; and time of fall, 111, 137, 138; instantaneous, 60; of motion, from its cause, 137; of motion, from its effect, 137

vera causa, 100, 180, 181; see also true cause

verification, 29, 31, 39, 41, 42, 64, 95, 101, 183, 186, 216, 221

via: inventionis, 98; *judicii*, 98; *nominalium*, 57; *realium*, 58; see also demonstration; discovery; nominalists; realists

Villoslada, R. G., 235, 265

virtus derelicta, 106

visual power, 35

vital spirits, 190

voluntarism, 54

vortex motion, 210

Wallace, W. A., 224, 257, 265–266

Wallis, C. G., 240, 255

Wartofsky, M. W., 212, 266

wave-particle construct, 3

Webster, C., 245, 266

weight, essential, 179

Weisheipl, J. A., 59, 62, 69, 217, 218, 222–226, 243, 266

Westfall, R. S., 247, 248, 266–267

Weyl, H., 211

wheel, rotating, 129, 137
Whewell, William (1794–
1866), 162, 205, 249
Whittaker, E., 223, 267
Whitteridge, G., 246, 267
Wicksteed, P. H., 215, 252
Wiener, P. P., 243, 267
Wilkie, J. S., 245, 267
William of Moerbeke (1215?–
1286), 46, 213–215
William of Sherwood (d.
1267?), 104
Willis, R., 245, 255
Wilson, C., 60, 224, 241, 267
Witelo (1225?–1278?), 222
worlds, plurality of, 113
world soul, 94, 154, 167
wound: circular, 13, 40, 77;

healing of, 33, 36, 39, 40,
77
Würschmidt, J., 229, 257
Wyclif, John (1330?–1384),
125

Zabarella, Jacopo (1533–1589),
118, 139, 144–149, 176,
232, 237, 238, 257; *De
methodis*, 238; *De motu
gravium et levium*, 178,
243; *De regressu*, 146, 147,
238; on the *Posterior Analytics*, 144, 148
Zilsel, E., 240, 267
Zimara, Marc Antonio (1460–
1532), 144